Study Learning Guide

for use with

Prepared by DEBORAH BOUTILIER
NIAGARA COLLEGE

Australia Canada Mexico Singapore Spain United Kingdom United States

Student Learning Guide for use with New Society, Fifth Edition

by Deborah Boutilier

Associate Vice President, Editorial Director:
Evelyn Veitch

Editor-in-Chief, Higher Education:
Anne Williams

Executive Editor:
Cara Yarzab

Marketing Manager:
Heather Leach

Senior Developmental Editor:
Kamilah Reid Burrell

Proofreader:
Rodney Rawlings

Manufacturing Coordinator:
Loretta Lee

Cover Design:
Johanna Liburd

Cover Image:
© Marshall Ikonography/Alamy

Printer:
Thomson West

ISBN-13: 978-0-17-610503-7
ISBN-10: 0-17-610503-4

Printed and bound in the United States
1 2 3 4 10 09 08 07

For more information contact Nelson, 1120 Birchmount Road, Toronto, Ontario, M1K 5G4. Or you can visit our Internet site at http://www.nelson.com

Preface

Welcome!

My name is Deborah Boutilier and I've had the distinct pleasure of writing this Student Learning Guide to accompany your very fine textbook. I've been teaching sociology for a really long time, and it's been my experience that students who use a learning guide usually get better grades than students who don't. That's one of the reasons I decided to take on this project. In addition to teaching sociology, I'm working on my doctoral degree—which is very cool because I get to identify with my students not only from the perspective of an instructor but also from the perspective of a student. You'll notice when you go through this guide that I usually write in the first person—almost as if I'm talking to you—and that's because I'm hoping that we'll connect as we go along, which will facilitate your learning process. Sometimes my ideas can be a little off the wall, but my students always tell me that that's what makes my classes so much fun and that's why they remember their material so well. Consequently, sometimes my chapter introductions read like stories or little vignettes as I share my life experiences both in and out of the classroom!

I really hope that you'll have some fun as you read the chapters in the textbook and then augment your learning with the exercises that I've created here. I've spent tons of time surfing the Internet to find the absolutely coolest links possible, and trust me, you could spend hours just checking out the additional information and trying out the interactive exercises that they contain. Some of them are a little controversial or a little challenging, so don't be afraid to push your limits and expand your mind—that's what learning is all about. Most importantly, don't be afraid to make mistakes—for all my book learning and fancy degrees, on some days I know that if I didn't learn from my mistakes, I wouldn't learn anything at all. So, when you do your tests, make sure that you go back to see where you went wrong *and* where you went right—it's an important part of the learning process.

How Can I Get the Most out of My Student Learning Guide?

This book is called a Student Learning Guide and not a Student Study Guide for some very specific reasons. Although it looks the same as other study guides that you might have used in the past, this supplemental learning tool is meant to act as a guide that will enhance your learning skills in the area of sociology, while providing the comprehensive understanding that you need to be successful in this course. Rather than merely testing your familiarity with the textbook or improving your memory retention, if used properly this guide will also give you the self-confidence and knowledge you will need to produce quality exams and research papers. More often than not, it is these important course requirements that determine your level of success or failure in any subject. Any student can study, but it's only the successful ones who can truly learn.

Your student learning guide is not too big or heavy to carry in your knapsack or briefcase. Instead, it's easy to access and convenient to use wherever or whenever you choose to learn. I've designed it so that you can learn from it with or without the use of your textbook; but it's important to remember that this guide should never be used as a substitute for your textbook, and that it works best in conjunction with your text. You'll notice that your learning guide mirrors the most important parts of the textbook, but also adds supplemental material to enhance your studying. From students I've taught over the years, I've found that bringing your learning guide to class and highlighting the topics that the instructor discusses during lectures is an excellent way to gain a greater understanding of the most important concepts that are likely to appear on a test or exam. Additionally, writing in your learning guide makes reviewing material for an exam much easier.

Your student learning guide is also simplistic in form, and it attempts, through a variety of ways, to guide your learning. You are ultimately responsible for actually accumulating and retaining educational knowledge; however, this book will help you to facilitate your learning so that it becomes something enjoyable, not stressful. I've been a student for most of my life and probably for more years than you've been alive, and I've tried to draw on some of the experiences that I've had as a student and instructor to help you have a greater understanding of how sociological theories really work in everyday life. I've had a totally awesome time creating this book, and I sincerely hope that you have just as much fun using it. Carry it with you in your knapsack and pull it out when you get a chance—in between classes, over morning coffee, or even while waiting for the bus. You'll be amazed at how much this little book can offer, but even more amazed at how much it can make you think!

Jump Start Your Brain!

When you get to this section, think of Bugs Bunny when he used to crack his knuckles before he played the piano! Or if that doesn't work, remember the *Seinfeld* episode where Jerry put the Tweety Bird Pez on Elaine's purse? In that one, they had made a bet about how professional pianists warm up before their performances, and sure enough, George's girlfriend told them that she just cracked her knuckles! This section will give you overall idea of the major concepts that you'll be learning in the chapter, and it's a good idea to take your time and really digest these points before you read more about them. Some of my students will go back and highlight these concepts once they've hit on them in the chapter to make sure that they've identified each of them. You can bet that your instructors will focus on these in their lectures, since they're the key issues the chapter deals with. In some cases, they will also act as guides your instructors might use when they create questions for exams or assignments.

True/False and Multiple Choice Questions

Objective-type questions not only test how much you remember from the textbook, they also make you think about what you've learned and what makes you remember what you've learned. From these questions, try to develop your learning skills and identify the tricks that you can use to help you retain material from the textbook. These questions are not designed simply to test your memory recall. These are tools that can help you learn. Your scores here should be a good indication of how well you understand each chapter. If your scores are not good (i.e., less than 60 percent), reread the chapter and try again. These questions are especially designed to help you prepare for any upcoming objective examinations and should act as good indicators of your potential performance levels.

Critical Thinking Questions

The study of sociology can be defined as the method by which ordinary people—like you and me—strive to understand what is happening to them and their society. Systematic observation of the structures, institutions, and forces in society encourages us to look beyond the strictly personal and individual events that affect our lives, allowing us to use our critical thinking skills to see the world from a different, more global perspective. The need for such understanding is urgent at this present historical moment when change is rapid and the future is uncertain. Because of these reasons and more, unprecedented opportunities to build a better world and improve the quality of life for many people lie ahead. You can capitalize on those opportunities, and the study of sociology can help. By taking this course and reading both your textbook and your student learning guide, you'll explore how society works, how it doesn't work, and the things that you can do to change it. By investigating sociological perspectives and developing a sociological imagination, you can gain the potential to change the world.

The critical thinking questions included at the end of each chapter require you to adopt a more subjective and critical attitude to the issues at hand. While it is important to learn about sociology, it is equally important to learn how to use the skills of a sociologist. It's great to grasp the meanings of the theories, but it's even better to be able to apply those theories to the real-life situations that confront us daily. Subjective learning gives us as much of an opportunity to enrich our understanding as objective learning, but it requires that we use a different set of skills. That's where the imagination part comes in. When you attempt to answer these questions, try to picture yourself in the situations, and don't be afraid to let your ideas come forth, regardless of how off the wall they may seem to you or others. Critical thinking is the key to understanding all of the different facets that one idea or theory can have. Let your imagination go crazy—you'll be surprised at what you come up with!

Time to Gear Down!

This is the section where you can really test your knowledge and summarize what you've learned from the chapter. Each summation represents a salient or really important point that was made in the chapter. What you might want to do here is just take a couple of deep breaths and relax. Then, slowly think about each summary point and see if you can remember where in the chapter that point was made. If you can't pinpoint it exactly, then you might want to back through the chapter and find it. If you're not up to doing all that work again, then you definitely want to remember that these summation points can act as key points for studying material. Don't even for a second think that they'll replace reading the chapter, because they won't—but they will give you a good idea of what to focus on if you're looking to go over the main themes in the chapter.

Web Links

If I must say so myself, I've done an excellent job of surfing the Net—for someone who can't even swim—and I have come up with some of the *coolest* links ever! Many of them will have much more information concerning main themes of the chapter, but some of are interactive, awesome, wild, and just plain weird.

Critical "Linking" Question

This final question asks you to investigate one of the Web Links in more detail—I've usually picked one that's especially neat—and will guarantee to get you Brownie points for discussion in class for sure!

Tips for Better Learning

1. Identify the kind of learner you are! Are you a visual, auditory, or kinaesthetic learner? Once you've identified your learning style (student services or counselling should be able to help with this), work on developing your reading, writing, and study skills. If your university or college offers workshops—attend them! Usually, the student services or student development centre offers classes in time management, relaxation, cxam anxiety, note taking, essay writing, and study skills. Though some may charge a nominal fee for these services, they are usually well worth the money and can often be invaluable to your education.
2. Use your library and all the talented people who work there. Most college and university libraries offer tours and information sessions that acquaint you with the many services they provide. If you don't know how to use the online databases, make that a priority on your to-do list. Researching from home can make your academic year—and career—so much better, and believe me, it's not that hard to do.
3. Get organized! Creating a schedule that you can live with often determines not only the success of your academic year, it also allows you to make the best of your leisure time as well. A social life is just as important as an academic life, so make sure that you include some recreational time in your planner. A good schedule is a realistic one. By scheduling time away from studies, you allow yourself to have some fun without worrying about upcoming assignments.
4. Get computer-literate and learn to type—learn how to use the Internet and the World Wide Web. This can cut your essay writing time in half. Most schools offer free two- or three-hour workshops on how to research on the Internet. I always advise my students to take a basic word-processing course as an elective credit in their degree program. That way not only do they earn credits they develop very crucial skills that will help them save time and money in the future—it's incredibly expensive to have someone else type a paper for you.
5. Improve your reading skills. Start out by replacing television with just a half-hour of reading. It doesn't matter what you read, as long as you get into the habit. Even

if it's recreational reading, it will enhance your reading skills, which will help improve the reading skills that you use for studying.

6. Work on your social skills and start networking now! By the time you graduate, companies will be looking for graduates who possess the people skills that will allow them to travel around the world and mingle with a diverse population. Additionally, we know that it's through networking that you might meet that acquaintance that may lead you to the job of your dreams.
7. Form a study group—many students find that this is a great way to study and reinforce information that they have learned. It also allows them to interact with each other and to get to know their classmates better.
8. Become a global thinker. The more you know about the world, the more information you'll have to apply in every class that you attend. It's incredibly important to have at least a bit of knowledge about global issues. You can easily do this by quickly reading the headlines or scanning the home page of your browser's news link.
9. Think outside the box. Although you might be majoring in business, if you get to take an elective, take something that completely unrelated to your field. Expand your mind across disciplines so that you become practiced in the art of cerebral flexibility. Employers really do look for this sort of thing when they hire.
10. Expand your researching skills. Knowledge is power, and it's time to start looking for that power in places you've not thought to look before. With the newer forms of media and technology coming out, take the time to learn how to access it for information, rather than just for fun. If you're on YouTube watching a funny video, stay on a bit longer and find a video that would also inform you on a realistic topic that you might be interested in.

A Word to the Educators Who Use This Guide

Having taught a variety of sociology courses over the past 15 years, I am well acquainted with the enormous amount of preparation time required to effectively present material that students will clearly understand and retain. This Student Learning Guide has been designed to provide thought-provoking material for students, as well as teachers. The objective-type questions make great quizzes, and the critical thinking questions will facilitate good group discussions and seminar presentations. In this edition, the Web Links at the end of each chapter are especially good and can provide for great seminar topics or assignments. They augment the chapters wonderfully, and the Critical Linking Question will for certain get the students actively involved.

About the Author

Deborah Boutilier teaches sociology in the School of General Studies at Niagara College. She holds a Master of Arts in Sociology from SUNYAB and a Masters of Education from Brock University. She is currently completing a doctoral degree in the department of

Sociology and Equity Studies at OISE/UT, where she is also employed as a graduate researcher.

Acknowledgments

I am forever grateful to Robert Brym—who continues to astound me with his wisdom and insight—and for the incredible talents of Kamilah Reid-Burrell at Thomson Nelson. I am very lucky to have had Dr. R. Boily, Lauralee Eldridge, and Sue Kowal play such an important role in this endeavour and in my life. And though I'm grateful for all the students that pass through my life, this work was inspired especially by three former students who have recently celebrated their own victories—Jess Dixon, Patrick Smith, and David Tyminski. Well done, gentlemen! Well done, indeed.

Contents

Chapter 1:	Introducing Sociology	1
Chapter 2:	Culture	9
Chapter 3:	Socialization	17
Chapter 4:	Gender and Sexuality	25
Chapter 5:	The Mass Media	33
Chapter 6:	Social Stratification	42
Chapter 7:	Gender Inequality: Economic and Political Aspects	51
Chapter 8:	Race and Ethnic Relations	60
Chapter 9:	Inequality Among Nations: Perspectives on Development	68
Chapter 10:	Families	77
Chapter 11:	Work and Occupations	87
Chapter 12:	Education	96
Chapter 13:	Religion	105
Chapter 14:	Deviance and Crime	113
Chapter 15:	Population and Urbanization	122
Chapter 16:	Sociology and the Environment	131
Chapter 17:	Health and Aging	140
Chapter 18:	Politics and Social Movements	148
Chapter 19:	Globalization	156
Chapter 20:	Research Methods	165
Chapter 21:	Networks, Groups, Bureaucracies, and Societies	173

Chapter 1
Introducing Sociology

Chapter Introduction

The goal of this chapter is to guide you to a place in your mind that will allow you to think and learn about sociology. There's a really good chance that you and many other students like you, have already "done" sociology, but have not been aware that your thoughts and behaviours were shaped by societal influences. How often do you question the "invisible" actions and reactions that surround the seemingly routine activity of meeting a friend for coffee? Do you ever wonder where the coffee beans came from or the daily wage of the people who picked them? Is having coffee with a friend simply about the process of drinking coffee or are there other, more important reasons for getting together? Does it matter if you have tea instead of coffee? In developing an understanding of how to think sociologically, you may often find yourself thinking and questioning things that you would usually take for granted. The process involved in thinking and questioning is almost always a good thing, and the study of sociology will make you feel better about trying to understand the complexities that surround the society in which you live.

Sometimes in the attempt to understand a new body of knowledge, you get so caught up in trying to determine what the subject is all about that you often fail to think about what the subject "is not" about. Brym's explanation of sociology offers a solid foundation on which to begin your understanding of this discipline by providing an excellent review of what sociology is and couching it within the differences that exist between sociology and other disciplines of investigation. By self-disclosing his own beginnings as a somewhat hesitant sociologist, Brym invites you to become a critical thinker as you delve into and investigate the social facts that make up *your* world!

Jump Start Your Brain!

In this chapter you will learn that:

1. The causes of human behaviour lie partly in the patterns of social relations that surround and penetrate people.
2. Sociologists examine the connection between personal troubles and social relations.
3. Sociological research is often motivated by the desire to improve the social world. At the same time, sociologists adopt scientific methods to test their ideas.
4. Sociology originated at the time of the Industrial Revolution. The founders of sociology diagnosed the massive social transformations of their day. They also suggested ways of overcoming the social problems created by the Industrial Revolution.

5. Today's postindustrial revolution similarly challenges us. The chief value of sociology is that it can help clarify the scope, direction, and significance of social change. Sociology can also suggest ways of managing change.

Quiz Questions

True or False?

1. Sociologists examine the connection between personal troubles and social relations. True or False
2. Because we are social beings that interact with others and the world around us, it is impossible to distinguish the social cause of phenomena from the physical and emotional causes. True or False
3. According to Durkheim, the Protestant belief that "religious doubts can be reduced and a state of grace assured if people work diligently" was called the Protestant work ethic. True or False
4. The feminist paradigm holds that male domination and female subordination are determined biologically. True or False
5. Symbolic interactionism stresses that people help to create their social circumstances and do not merely react to them. True or False
6. Durkheim argued that suicide rates vary due to differences in the degree of social solidarity in different groups. True or False
7. Because it is hard to study large groups in society, sociologists must also study individuals. True or False
8. It is believed that Jews are less likely to commit suicide than Christians because centuries of persecution have turned them into a group that is more defensive and tightly knit. True or False
9. Sociologists call relatively stable patterns of social relations "social structures." True or False
10. By stressing the importance and validity of subjective meanings, conflict theorists also increase respect for and tolerance of minority and deviant viewpoints. True or False

Multiple Choice

1. Which of the following is *not* true?

    a. Sociologists examine the connection between personal troubles and social relations.

    b. Sociology originated at the time of the Age of Enlightenment

    c. The causes of human behaviour lie partly in the patterns of social relations that surround people.

   d. Sociological research is often motivated by the desire to improve the social world.

2. At the end of the nineteenth century, ____________ demonstrated that suicide is more than just an individual act of depression.

   a. Marx

   b. Engels

   c. Durkheim

   d. Weber

3. ____________ structures lie outside and above the national level.

   a. International

   b. Macro

   c. Global

   d. Micro

4. This chapter showed that diminished ____________ relations have powerfully influenced suicide rates.

   a. psychological

   b. social

   c. religious

   d. intimate

5. A ___________ is a tentative explanation of some aspect of social life that states how and why certain facts are related.

   a. sociological question

   b. theory

   c. null hypothesis

   d. statement

6. Compared to the youth of the 1960s, young people today are ____________ likely to take their own lives if they happen to find themselves in the midst of a personal crisis.

   a. more

   b. less

   c. not at all

   d. equally

7. Half a century ago, the great American sociologist ____________ called the ability to see the connection between personal troubles and social structures the "sociological imagination."

   a. Karl Marx
   b. Emile Durkheim
   c. Max Weber
   d. C. Wright Mills

8. The term sociology was coined by the French social thinker _____________.
   a. Auguste Comte
   b. Emile Durkheim
   c. Talcott Parsons
   d. Jean Jacques Rouseau

9. Which of the following is *not* a commonly accepted basis for knowing that something is "true" in our everyday lives?
   a. qualification
   b. overgeneralization
   c. internal logic
   d. ego-defence

10. _____________ are ideas about what is right and wrong.
   a. Norms
   b. Laws
   c. Folkways
   d. Values

11. ____________ suggests that reestablishing equilibrium in society can best solve most social problems.
   a. Feminism
   b. Symbolic interactionism
   c. Conflict theory
   d. Functionalism

12. Class conflict lies at the centre of _____________'s ideas.
   a. Karl Marx
   b. Emile Durkheim
   c. George Herbert Mead
   d. John Porter

13. _____________ refers to the technology-driven shift from manufacturing to service industries and the consequences of that shift on virtually all human activities.

a. The Industrial Revolution

b. The Technological Revolution

c. The Postindustrial Revolution

d. Alvin Toffler's *Future Shock*

14. "The proper place for women is in the home. That's the way it's always been." This statement represents knowledge based on____________.

a. casual observation

b. overgeneralization

c. tradition

d. selective observation

15. What does Yorick's dilemma tell us?

a. That hard work pays off.

b. That life is finite.

c. That people live on in memories, after their death.

d. That true friendship never dies.

Critical Thinking Questions

1. Durkheim's study of suicide in France remains a landmark work in the field of sociology. Often touted as the supremely antisocial and nonsocial act, suicide is thought to be a highly individual act that occurs outside the realm of social forces. When people that we know or celebrities take their own life, our first reaction is to try and understand why. How many times have you tried to determine the motive behind a suicide? Have you ever thought that suicide is inextricably linked to social forces? Though many theories exist, will we ever discover the story behind Kurt Cobain's tragic death? Why would he choose to end his life at a time when it seemed as though his life was perfect? What are the social factors that you think contributed to his death?
2. A good understanding of the fundamental theories is an absolute necessity that can often determine your success in any sociology course. For this exercise, consider the existence of prostitution in society. Remember that it is legal in some countries but illegal others. How would a structural functionalist, a conflict theorist, a feminist, and a symbolic interactionist explain prostitution?
3. According to the work of Mark Granovetter (1973), you are likely to find a job faster if you understand the "strength of weak ties" in microstructural settings. Do you agree with his findings? Do you know people who got great jobs because of their "connections"? How helpful is it to know someone when it comes to finding meaningful work?
4. Sociology is becoming increasingly reliant on the knowledge-worker, and the ability to acquire, retain, and apply a variety of information is an important tool

for anyone that plans to enter the workforce. Since you're using this learning guide, you are probably enrolled at a college or university that will help you with the processes that are involved in understanding and retaining knowledge. To what extent do you think that your time spent in a formal educational setting will assist you in obtaining the career of your dreams? Examine your reasons for being in school. How will you feel if, after you graduate, you don't get the job that you've spent so much time and money preparing for?

Time to Gear Down!

At the conclusion of this chapter, you should be able to discuss or write about the following without having to rely on the textbook:

1. Durkheim showed that even apparently nonsocial and antisocial actions are influenced by social structures. Specifically, he showed how levels of social solidarity affect suicide rates.
2. Due to the rise in youth suicide, the pattern of suicide rates in Canada today is not exactly the same as in Durkheim's France. Nevertheless, Durkheim's theory explains the contemporary Canadian pattern well.
3. Sociologists analyze the connection between personal troubles and social structures.
4. Sociologists analyze the influence of three levels of social structure on human action: microstructures, macrostructures, and global structures.
5. Values and theories suggest which sociological research questions are worth asking and how the parts of society fit together. A theory is a tentative explanation of some aspect of social life. It states how and why specific facts are connected. Research is the process of carefully observing social reality to assess the validity of a theory.There are four major theoretical traditions in sociology. Functionalism analyzes how social order is supported by macrostructures. The conflict paradigm analyzes how social inequality is maintained and challenged. Symbolic interactionism analyzes how meaning is created when people communicate in microlevel settings. Feminism focuses on the social sources of patriarchy in both macro and micro settings.

   The rise of sociology was stimulated by the scientific, industrial, and democratic revolutions.
6. The Postindustrial Revolution is the technology-driven shift from manufacturing to service industries and the consequences of that shift for virtually all human activities.
7. The causes and consequences of postindustrialism form the great sociological puzzle of our time. The tension between autonomy and constraint, prosperity and inequality, and diversity and uniformity are among the chief interests of sociology today.

Web Links

Journal of Mundane Behaviour

http://www.mundanebehavior.org/index2.htm

The *Journal of Mundane Behaviour* "was a blind peer-reviewed scholarly and publicly-oriented journal devoted to the study of the 'unmarked'—those aspects of our everyday lives that typically go unnoticed by us, both as academics and as everyday individuals"—the very same thing that I wrote about in the introduction of this chapter! Check out some of the "out there" topics that are sociologically and scientifically presented in a way that makes you think like you never have before. Though now defunct, this appears to be weird and wild scientific sociology at its finest.

Social Studies

http://www.ncss.org

This website, produced by the National Council for Social Studies, offers both educators and students a variety of information from the world of social studies. It's geared mainly toward teachers, and if you investigate the resources site, it's neat to see the lessons plans that teachers would use to teach children about the world around us. A totally different way of looking at things.

Electronic Journal of Sociology

http://www.sociology.org

An incredibly valuable free website that offers you access to some excellent online sociological journals, as well as some instructional material should you ever think about publishing!

Directory of Social Science

http://dir.yahoo.com/Social_Science/Sociology

Make sure to add this one to your list of favourites—full access to virtually every topic imaginable in the quest for sociological information.

A Sociological Tour Through CyberSpace

http://www.trinity.edu/mkearl/index.html#in

A Sociological Tour through CyberSpace offers you virtually (get it?) everything you ever wanted to know about sociology but didn't know what to ask. Literally, you could spend weeks on this site, discovering things you don't even want to know, but in reality, it's an excellent tool for studying and adding additional information to your lectures and notes. Make sure to spend some time learning how to navigate this site properly!

Critical "Linking" Question

Go to the *Journal of Mundane Behaviour* website and click on the "Outburst" link or from the drop-down menu, check out the Outburst Archives. Some of the past rants are unbelievable. If you were going to submit an article for the Outburst section of this journal, what would your topic be?

Solutions

True or False?

1. T
2. F
3. F
4. F
5. T
6. T
7. F
8. T
9. T
10. F

Multiple Choice

1. B
2. C
3. C
4. C
5. B
6. A
7. D
8. A
9. C
10. D
11. D
12. A
13. C
14. C
15. B

Chapter 2
Culture

Chapter Introduction

Stop for a moment and reflect on how you felt about living in Canada before you read this chapter and compare it with how you feel now. Have your feelings about your homeland changed in any way? Do you gct a better sense of just how truly diverse our country is? I often wonder what it would be like to travel around the world, and have always been envious of those who do spend a lot of time travelling and experiencing new and exciting places. I don't know about you, but I've never had the "travel bug" and am quite content to stay right where I am. Reading this chapter allows you to visit a number of different cultures without having to leave the comfort of your own home and enlightens you on some of the more positive and negative consequences of living in a culturally diverse community. More than ever, Canadian students are travelling to China, Japan, and Korea to teach English as a second language to students of all ages. Practise your newfound critical thinking skills and imagine what your life would be like if you had to live in a foreign country for a year. How would you prepare yourself for the changes that you would face? What would be the determining factor in choosing a country. Would the weather affect your decision more than the food? Would the politics of that nation require you to alter some of your current political beliefs? So many times we are faced with opportunities that challenge us in a variety of ways. Though these challenges often mean having to make important decisions, can you imagine living in a world without diversity? How very boring that would be!

This chapter should make you think about Canada from a number of different viewpoints and hopefully you'll see how cultural sharing takes place through social transmission. You'll soon learn that culture is the sum of the socially transmitted ideas, practices, and material objects that enable people to adapt to and thrive in their environments.

Jump Start Your Brain

In this chapter you will learn that:

1. Culture is the sum of ideas, practices, and material objects that people create to adapt to, and thrive in, their environments.
2. Humans have been able to thrive in their environments because of their unique ability to think abstractly, cooperate with one another, and make tools.
3. We can see the contours of culture most sharply if we are neither too deeply immersed in it nor too much removed from it.
4. As societies become more complex, culture becomes more diversified and consensus declines in many areas of life. This increases human freedom.

5. As societies become more complex, the limits within which freedom may increase become more rigid. This constrains human freedom.
6. Although culture is created to solve human problems, it sometimes has negatives consequences that create new problems.

Quiz Questions

True or False?

1. As societies become more complex, culture becomes more diversified and consensus declines in many areas of life. This decreases human freedom. True or False
2. Rewards and punishments aimed at ensuring conformity are known as sanctions of the system of social control. True or False
3. Culture is the sum of ideas, practices, and material objects that people create in order to adapt to and thrive in their world. True or False
4. Judging another culture exclusively by the standards of one's own is called ethnocentrism. True or False
5. Inexpensive international travel and communication make contacts between people from diverse cultures routine. True or False
6. Although culture is created to solve human problems, it sometimes has negative consequences that create new problems. True or False
7. The postmodern condition disempowers ordinary people, leaving them with little or no control over their own fate. True or False
8. As societies become more complex, the limits within which freedom may increase become more rigid, constraining human freedom. True or False
9. The English language is dominant because Britain and the United States have been the world's most powerful and influential countries—economically, militarily, and culturally—for 200 years. True or False
10. The rights revolution is the process by which new Canadians obtain equal rights under the law and in practice. True or False

Multiple Choice

1. Globalization destroys _____________ isolation, bringing people together in what Marshall McLuhan called a global village.

   a. political, economic, and cultural

   b. cultural, technological, and religious

   c. political, judicial, and religious

   d. cultural, political, and judicial

2. Which of the following is *not* true?

a. Postmodernism involves an eclectic mixing of elements from different times and places.

b. Postmodernism involves the erosion of authority.

c. Postmodernism involves a heightened awareness of the masses.

d. Postmodernism involves the decline of consensus about core values.

3. Which is not a tool in the human survival kit?

a. abstraction

b. cooperation

c. production

d. technology

4. Sociologists define ____________ as all the ideas, practices, and material objects that people create to deal with real-life problems.

a. culture

b. society

c. postmodernism

d. subculture

5. Research shows that the average North American prefers to stand ____________ inches away from strangers or acquaintances when they are engaged in face-to-face interaction.

a. 20 to 26

b. 30 to 36

c. 40 to 46

d. 50 to 56

6. Cultural freedom develops within definite limits. In particular, our lives are increasingly governed by the twin forces of ____________.

a. consumerism and rationalization

b. consumerism and technology

c. technology and rationalization

d. none of the above

7. A __________ is a cultural ceremony that marks the transition from one stage of life to another or from life to death.

a. ritual

b. custom

c. rite of passage

   d. tradition

8. Clocks, known as *Werkglocken*, were created in Germany and signified .
   a. the creation of unions
   b. the practice known as "work to rule"
   c. the beginning of the workday, the timing of meals, and quitting time
   d. the efficiency of assembly-line work
9. ____________ is the application of the most efficient means to achieve given goals and the unintended, negative consequences of doing so.
   a. Consumerism
   b. Bureaucracy
   c. Taylorism
   d. Rationalization
10. The capacity to create ideas or ways of thinking is referred to as ____________.
    a. production
    b. cooperation
    c. abstraction
    d. creativity
11. The tendency to define ourselves in terms of the goods we purchase is known as ____________.
    a. Instant Gratification Syndrome
    b. consumerism
    c. competition
    d. Fashionism
12. As societies become more complex, culture becomes more diversified and consensus declines in many areas of life. This also increases ____________.
    a. our ability to become better consumers
    b. our human freedom
    c. divisions among the classes that exist in society
    d. our awareness of other cultures
13. Allowing clocks to regulate our activities precisely seems the most natural thing in the world and is a pretty good sign that the internalized *Werkglock* is, in fact, a product of ____________.
    a. capitalism
    b. culture

c. technology

d. changing times

14. _______________ is the godfather of heavy metal. Beginning in the 1960s, he and his band, Black Sabbath, inspired Metallica, Kiss, Judas Priest, Marilyn Manson, and others to play loud, nihilistic music, reject conventional morality, embrace death and violence, and foment youthful rebellion and parental panic.

a. Tommy Lee

b. Perry Como

c. Glen Campbell

d. Ozzy Ozbourne

15. Hockey legend Wayne Gretzky would tuck only the right side of his jersey into his pants. This superstitious practice helped to put his mind at ease before and during play. It is an example of how people create _____________ to cope with anxiety and other concrete problems they face.

a. culture

b. safety nets

c. personal anomalies

d. routines

Critical Thinking

1. Do you think there is any foundation for superstitious beliefs? If not, how can you explain the fact that so many people are superstitious? Do you have any superstitions? Have you ever defied a superstition—walked under a ladder or broken a mirror? If so, what were the results?
2. Have you ever lived in another culture? If so how does your own culture differ from the one you lived in or visited? If you haven't lived in another culture, you likely know someone who comes from another country. Have you learned about their customs and beliefs? To what degree does ethnocentrism affect how and what you think of that person?
3. Can you imagine buying a Big Mac from a vending machine? I think that Morgan Spurlock's movie *Super Size Me* has changed the way that many people feel about fast food, but in this chapter Ritzer paints a fairly bleak picture of "Mcjobs" and ways of buying food. Is it really as bad as he makes it seem? If it is, what accounts for the fact that teenagers work in most fast-food restaurants? What are other employment opportunities for teens?
4. "It's not the steak we sell. It's the sizzle." This statement was made about advertising in the 1940s. How much does advertising really influence us? At what age do you think advertising starts to influence us? Do we ever become aware of its influences? Have you bought or boycotted a product just because of the advertising? Are you likely to do so, and under what circumstances?

Time to Gear Down!

At the conclusion of this chapter, you should be able to discuss or write about the following, without having to rely on the textbook:

1. Humans have been able to adapt to their environments because they're able to create culture. Specifically, they can create symbols, cooperate with others, and make tools that enable them to thrive.
2. Culture can be invisible if we are too deeply immersed in it. The cultures of others can seem inscrutable if we view them exclusively from the perspective of our own culture. Therefore, the best vantage point for analyzing culture is on the margins, as it were—neither too deeply immersed in it nor too much removed from it.
3. Culture becomes more diversified and consensus declines in many areas of life as societies become more complex. This increases human freedom, giving people more choice in their ethnic, religious, sexual, and other identities.
4. So much cultural diversification and reconfiguration has taken place that some sociologists characterize the culture of our times as postmodern. Postmodernism involves an eclectic mixture of cultural elements from different times and places, the erosion of authority and the decline of consensus around core values.
5. Underlying cultural diversification is the rights revolution, the process by which socially excluded groups have struggled to win equal rights under the law and in practice.
6. Although the diversification of culture increases human freedom, the growth of complex societies also established definite limits within which diversification may occur. This is illustrated by the process of rationalization (the optimization of means to achieve given ends) and the growth of consumerism (which involves defining one's self in terms of the goods one purchases).

Web Links

Smithsonian Institution

http://www.si.edu

This is simply an amazing website. The Smithsonian Institution offers an incredible abundance of information from cultures around the world. Time spent learning from this site is definitely worthwhile.

Canadian Heritage—International Relations and Policy Development

http://www.canadianheritage.gc.ca/progs/ai-ia/ridp-irpd/02/index_e.cfm

This extremely comprehensive site both explains and supports Canada's commitment to cultural diversity.

Symbols and Signs

http://www.symbols.net

This is a wickedly cool website that you can totally get lost in! Symbols form the basis for communication and language is the most important set of symbols that we know. You'll be astounded when you start checking out all the other symbols and signs that you recognize on this site!

The Global Village

http://archives.radio-canada.ca/400d.asp?id=1-74-342-1814

Check out this video clip of Marshall McLuhan as he first discusses what would become an internationally known catchphrase that captures the inception of cultural diversity and globalization.

The Billboard Liberation Front

http://www.billboardliberation.com/index.php

A group of highly secretive professionals hijack topnotch advertising billboards and tweak them to reflect a more realistic point of view—all without getting caught! Truly, truly amazing.

Critical "Linking" Question

If you go to the **Billboard Liberation Front** website and click on their Mission link, it explicitly outlines *exactly* how you would go about completely transforming an existing billboard. If you went to a website that told you how to create a nuclear weapons device, you're liable to end up on the FBI's Most Wanted list. It seems that vigilante justice exists in the advertising industry! Can you think of a billboard that you'd like to redesign?

Solutions

True or False?

1. F
2. T
3. T
4. T
5. T
6. T
7. F
8. T
9. T
10. F

Multiple Choice

1. A
2. C
3. D
4. A
5. B
6. A
7. C
8. C
9. D
10. C
11. B
12. B

13. A
14. D
15. A

Chapter 3
Socialization

Chapter Introduction

When does socialization begin and when does it end? Is it something that we have control over or something that occurs without conscious thought? Can we become better people through socialization? Answers to these questions have invariably remained the same over time, and I think that it's because we take the process of socialization for granted. We don't often think of changing the way we've been socialized because we're really not aware that it has happened until after the fact. This chapter explains that socialization is a lifelong process that begins when you're born and doesn't end until your life is over, however, I often argue that the process of socialization begins even before you're born. Think of the friends or family that you know that have had children. What are the preconceived notions that people have about newborn babies? What colour will the nursery be if it's a girl? What if it's a boy? Although these gender stereotypes are not as prevalent as they used to be, I think that even before they're born, babies have started the socialization process.

Socialization is all about the relationships and circumstances that you use to explore who you are and what you are about. Sociologists and psychologists agree that your self-image really depends a lot on social interaction. Since we spend the majority of our formative years with relatives, the family is the most important agent of socialization. As a general rule, our family members have more influence over our lives than our friends and peers even though it may not seem like it. Other agents of socialization include the media and religion, but the degree to which these institutions affect the definition of our social selves is largely an individual determination. I tend to think that I can't be swayed toward the purchase of a certain product simply by its advertising campaigns, but then there are other times when I know that I buy certain things based solely on their advertising. When younger, I often followed the advice of my peers rather than the advice of my parents and admittedly, more than once, make some very unwise decisions. Though we are solely responsible for defining who we are, we ultimately make this decision through interacting with others.

This chapter takes the approach that socialization is an active process, one in which those being socialized participate in and contribute to that socialization. The authors also point out that it's an interactive process in which those who are socializing are undergoing a learning process themselves. What I think that you'll learn is that the most important learning, the crucial learning that transforms us into cultural beings, proceeds through interaction with others who are important to us and, usually, to whom we are also important.

Jump Start Your Brain!

In this chapter you will learn that:

1. Socialization refers to the processes that allow people to become members of society, develop a sense of self, and learn to participate in social relationships with others.
2. Socialization takes place at all stages of the life cycle and in a variety of settings: families, schools, peer groups, the mass media, and occupational groups.
3. Among the major contributors to socialization theory are Charles Cooley (who argued that individuals develop a sense of self as they interact with others), George Herbert Mead (who focused on the way we actively create a sense of self by taking the roles of others), and Paul Willis (who demonstrated that young people creatively participate in generating and maintaining their sense of self, both in accordance with and sometimes in opposition to their social context).
4. Socialization continues into adulthood. As people work, marry, divorce, raise children, and retire, they enter new relationships with others, learn new behaviour, and adopt new roles.
5. Sometimes our self-concept undergoes abrupt change as we learn new role identities and negotiate a new self-image. Such "resocialization" occurs when we replace our way of life with a radically different one. It is most evident in jails, mental hospitals, and boot camps, and in religious and political conversions.

Quiz Questions

True or False?

1. Sociologists are suspicious of explanations that emphasize biological inheritance because such explanations often shift from an initial focus on individual differences to an emphasis on group differences. True or False
2. Sociologists and psychologists both agree that even though we are influenced by others, our self-image is a creation based solely on inner reflection. True or False
3. Taking the role of the other is an essential skill that a child must develop to be an effective member of society. True or False
4. George Herbert Mead's dramaturgical approach uses the analogy of life as theatre and of people as actors putting on performances for one another. True or False
5. Socialization can be a matter of life and death. True or False
6. The peer group is the only agent of childhood socialization over which adults have little control. True or False
7. Studies show that gender stereotyping has decreased but is still prevalent in children's literature. True or False
8. A peer group is a set of individuals who are about the same age, share similar interests, and enjoy a similar social status. True or False

9. The consensus among sociologists is that violence in the mass media may push some people to engage in more violent acts. True or False
10. Becoming parents accentuates gender roles and the sexual division of labour in heterosexual couples. True or False

Multiple Choice

1. Mead used the term _____________ to refer to the widespread and shared set of cultural norms and values used in self-evaluation and developing concepts of the self.

   a. taking the role of the other

   b. the generalized other

   c. the looking-glass self

   d. self-fulfilling prophecy

2. ____________ is the process by which aspirants to a particular social role being to discern what it will be like to function in that position.

   a. Socialization

   b. Adult socialization

   c. Anticipatory socialization

   d. Primary socialization

3. The self, for Cooley, has three major elements. Which of the following is *not* one of them?

   a. how we think our physical appearance is seen by another person

   b. our perception of other people's judgments about us

   c. our actual appearance

   d. our reaction to how we believe others judge us

4. ____________ uses the individual rather than the group as the frame of reference; emphasis is placed on the development of the person.

   a. Ethnomethodology

   b. Symbolic interactionism

   c. Structural functionalism

   d. Radical theorism

5. A ___________ is a group of people within the larger culture that has distinctive values, norms, and practices.

   a. sect

   b. counterculture

   c. cult

d. subculture

6. Rosenthal and Jacobson (1968) demonstrated the power of ____________ when an expectation leads to behaviour that causes the expectation to become a reality.

a. the looking-glass self

b. intrinsic predetermination

c. the self-fulfilling prophecy

d. subconscious thinking

7. Adolescence as a distinct period of life is a product of ______________.

a. socialization

b. dual-parent employment

c. industrialization

d. war

8. Which of the following is *not* a characteristic of Freud's theory?

a. gender

b. id

c. superego

d. ego

9. When we share in both the larger society and in a specific part of it, we are influenced by distinctive____________ of family, friends, class, and religion.

a. primary groups

b. peer groups

c. subcultures

d. relations

10. Which of the following describes groups of people usually of a similar age and equal social status?

a. cohorts

b. subcultures

c. peer groups

d. formal groups

11. Within ____________ of childbirth, first-time parents saw daughters as softer, finer-featured, and more delicate than sons.

a. two days

b. one day

c. a few hours

d. an hour

12. Conversion is often an example of ____________, the process of discarding former patterns of behaviour and belief and accepting new ones, although at times, reluctantly.

    a. resocialization

    b. anticipatory socialization

    c. brainwashing

    d. primary socialization

13. For most Canadians, mandatory retirement from the labour force occurs at age ____________.

    a. 55

    b. 60

    c. 65

    d. 70

14. Of all the functions of ____________, adjusting children to a social order—which offers a preview of what will be expected of them as they negotiate their way to adulthood—may be the most important.

    a. a peer group

    b. the family

    c. the media

    d. education

15. It is during ____________ that most dramatic transformations of identity, status, and social relationships tend to occur.

    a. infancy

    b. childhood

    c. adolescence

    d. adulthood

Critical Thinking

1. The nature vs. nurture debate has been around for many years, but it is still controversial. As we try to understand social change, we often fall back to this age old argument. Does nature play a bigger role than the environment in determining who we are, or is it the other way around? What do you think? Use this argument to discuss intelligence and then use it to discuss alcoholism—do your answers differ?

2. The self-fulfilling prophecy is a frequently debated topic in sociology, and Rosenthal and Jacobson's IQ study serves as a landmark case in the study of

socialization. They conducted their research in 1968. Do you consider this study to be ethical? Do you think that you could replicate this study today? If so, what changes would you make to the methodology of this work?

3. To be socialized means to learn how to act and interact appropriately with others and to transform oneself into a member of society. To be socialized is also to develop a self, a sense of individual identity that allows us to understand ourselves and differentiate ourselves from others. In this chapter, Paul Willis argues that we are all creative individuals trying to transform the world in ways that allow us to express and control our selves. Take a minute and think of the different ways in which you make your "mark" in and on this world. The authors of your textbook use E-mail addresses as an example—what others can you think of?

4. Peer pressure is an important agent of socialization and with increased levels of computer literacy and online "messaging," peer groups today are more extended and powerful than ever. Try to be honest with yourself as you answer if you've ever crumbled and submitted under peer pressure. If so, what have you done? Any regrets? Any permanent damage?

Time to Gear Down!

At the conclusion of this chapter, you should be able to discuss or write about the following, without having to rely on the textbook:

1. Socialization is an active process through which human beings become members of society, develop a sense of self, and learn to participate in social relationships with others.

2. Each of us is born with a set of human potentials. Nature and nurture interact in contributing to human development.

3. Socialization is lifelong, typically involving relationships with family, school, peer groups, mass media, and occupational groups. Ours is an age-graded society as well, and early childhood, adolescence, adulthood, and old age or retirement are significant stages; different roles and responsibilities are associated with each stage.

4. Because of its importance, many scholars have focused on examining socialization as an active, interactional process. Charles Horton Cooley was noteworthy for his concept of the "looking-glass self," which stressed that we view ourselves as we think others view us. George Herbert Mead emphasized how people assume roles by imagining themselves in the roles of others. Paul Willis, looking at the ways in which young people do not merely accept the world around them but transform them into a symbolic expression of their particular identity and their meaningful culture, has shown the intimate links connecting the acting individual and the broader social context.

5. Gender socialization is the learning of masculine and feminine behaviour roles. From birth and in every area of social life, the socialization of the sexes in terms of content and expectations makes the socially constructed gender role more

significant than the biological role of male or female. Assumptions about appropriate male and female attributes limit the range of acceptable behaviour and options for both sexes.

6. The most important agent of socialization is the family. As the examples of orphanages and child neglect demonstrate, initial warmth and nurturing are essential to healthy development. The self-concept formed during childhood has lasting consequences.
7. The central function of schools in industrial society is the teaching of skills and knowledge, but they also transmit society's central cultural values and ideologies. Schools expose children to situations in which the same rules, regulations and authority patterns apply to everyone.
8. Peer groups provide young people with a looking glass unclouded by love or duty, and an opportunity to learn roles and values that adults do not teach.
9. The traditional mass media are impersonal and large-scale socializers. New forms of media are more interactive and allow people to play with and try out different identities.
10. During adulthood, individuals are socialized as they get jobs, marry, divorce, raise children, retire, and prepare for death. All of these roles involve new and different relationships with others and also act as guidelines for behaviour.
11. Sometimes there are abrupt changes in our self-concept, and we must learn new role identities and negotiate a new self-image. Resocialization occurs when we abandon or are forced to abandon our way of life and self-concept for a radically different one. This is most efficiently done in total institutions—for example, jails, mental hospitals, and boot camps—or in religious or political conversions.

Web Links

The Genome Project Decodes the Nature vs. Nurture Debate

http://www.pbs.org/wgbh/nova/genome/debate.html

You can either surf and read or actually watch video clips of the nature vs. nurture debate as scientists uncrack the DNA code on this NOVA special.

Homeschooling

http://www.canteach.ca/links/linkhomeschool.html

This is a great site to learn more about homeschooling It provides information on the advantages and disadvantages of homeschooling and other related topics.

Beating Peer Pressure

http://teenadvice.about.com/cs/peerpressure/a/blpeerpressure.htm

No one likes to admit it, but we've all crumbled under peer pressure at one time or another. I always ask my students what they've done and I'm never surprised by the answers—OK, sometimes I am, but I try not to show it. Check out this website to see what you can do to avoid the pitfalls of peer pressure!

Queer Philosophy Discussion Forum

http://www.queerphilosophy.com/?q=node

This open forum website discussion group looks like it's well monitored, extremely objective and most importantly, respectful. If you've got an inkling to discuss queer issues and philosophies—or just have some questions you'd like answers to—this might be a site you'd like to investigate. It looks as though they're looking for volunteers on site maintenance as well.

Knowledge Socialization

http://www.research.ibm.com/knowsoc/project_index.html

A very unique and interesting way of looking at the socialization process as it happens through storytelling. This project uses it as it applies to organizations.

Critical "Linking" Question

Check out the Homeschooling website and click on any of links for more information. If you had the choice (and you had no financial concerns) would you homeschool your children?

Solutions

True or False?

1. T
2. F
3. T
4. F
5. T
6. T
7. T
8. T
9. F
10. T

Multiple Choice

1. B
2. C
3. C
4. B
5. D
6. C
7. C
8. A
9. C
10. C
11. B
12. A
13. C
14. D
15. C

Chapter 4
Gender and Sexuality

Chapter Introduction

In today's society men, women, and children are becoming more and more aware of their sexuality and the sexuality of others. Unfortunately, this doesn't always mean that everyone has become more accepting and tolerant of sexual difference or diversity. Although we know that Canadians are becoming more tolerant of homosexuality and same sex marriage, we also know that homophobia and stereotypical beliefs surrounding homosexuality are still prevalent in today's world.

The story of David Reimer acts as an incredible reminder of the importance of enjoying a quality of life that meets our own expectations. Is living inside a shell that masks our inner beings really living? How can we enjoy each day knowing that there are precious moments of our lives that we're not living to the fullest? Sexual acceptance and tolerance of others are such incredibly important concepts to both understand and practise. As our world grows more and more culturally diverse, it also expands its boundaries in terms of social and sexual acceptance for all—not just those who fit the skin that they're born in, but also those who struggle to get out.

This chapter asks how we define what is male and what is female and investigates the relationship between biological sex and the attitudes and behaviours that we associate with being male or female. What are the implications of this relationship for our sexual identity and sexual relationships? While introducing you to the concepts of sex and gender, this chapter also offers a wealth of sociological and psychological evidence that explains how the interpretation of these two terms can affect your life and the lives of others.

Jump Start Your Brain!

In this chapter you will learn that:

1. Sex refers to biological differences between males and females, while gender refers to the attitudes, beliefs, and behaviours we associate with masculinity and femininity.
2. Individuals form a "gender identity" or a sense of biological, psychological, and social belonging to a particular sex. Individuals also learn to play a "gender role," that is, to act in accordance with expectations about how members of their gender are supposed to behave.
3. Sexuality refers to activities intended to lead to erotic arousal and produce a genital response. Sexuality is guided by a set of social "scripts" that tell us whom we should find attractive, when and where it is appropriate to be aroused, when it is permissible to have sex, and so forth.

4. There are two major perspectives on the relationship among sex, gender and sexuality. One ("essentialism") holds that gender roles and sexual scripts develop naturally from biological differences between the sexes. The other ("social constructionism) holds that gender roles and sexual scripts emerge in response to the different social positions men and women carry.
5. Because of the way gender and sexuality are structured in our society, intolerance for sexual minorities is widespread and male sexual aggression against women is common.
6. A substantial decrease in gender inequality is now possible. The redefinition of sexuality is an important step in that process.

Quiz Questions

True or False?

1. Your gender depends on whether you were born with distinct male or female genitalia and a genetic program that released either male or female hormones to stimulate the development of your reproductive system. True or False
2. Transgendered people are individuals who want to alter their gender by changing their appearance or resorting to medical intervention. True or False
3. Sociobiologists are concerned with male female differences in sexual scripts, the division of labour at home and in the workplace, mate selection, sexual aggression, and so forth. True or False
4. Sexual pluralists believe that pornography does not have to reinforce the domination and degradation of women by men but it can allow women to create their own sexual fantasies. True or False
5. Premarital sex is widely accepted by the Canadian public. True or False
6. Essentialists argue that gender differences are not the product of biological properties, whether chromosomal, gonadal, or hormonal. Instead, gender and sexuality are products of social structure and culture. True or False
7. Research demonstrates that, in general, victims of sexual assault are selected less because of sexual desirability than because of their availability and powerlessness. True or False
8. Quid pro quo sexual harassment involves sexual jokes, comments, and touching that interferes with work or creates an unfriendly work setting. True or False
9. North Americans' expectations about how men and women are supposed to act have not changed much over the last 40 years. True or False
10. In our society, there is little formal socialization—that is, systematic instruction—regarding sexuality. True or False

Multiple Choice

1. _______________ refers to the way a person derives sexual pleasure, including whether desirable partners are of the same or a different sex.

   a. Gender orientation

   b. Gender identity

   c. Bisexual

   d. Sexual orientation

2. A 1997 survey shows that _____________ percent of North American women want to lose weight.

   a. 22

   b. 56

   c. 77

   d. 89

3. At the point of conception, a newly formed zygote has _____________ chromosomes. If the last chromosome has an XX pattern, the zygote becomes a female.

   a. 24

   b. 36

   c. 46

   d. 48

4. _____________ people are individuals who want to alter their gender by changing their appearance or resorting to medical intervention.

   a. Asexual

   b. Transvestic

   c. Hermaphroditic

   d. Transgendered

5. Differences between the sex organs are noticeable by the _____________ week after conception.

   a. fifth

   b. fourteenth

   c. twentieth

   d. thirty-second

6. One's identification with, or sense of belonging to, a particular sex—biologically, psychologically, and socially—is known as one's .

a. gender identity

b. gender

c. sex

d. sexual orientation

7. In the agricultural era, economic production was organized around the _____________.

a. household

b. father's income

c. factory

d. skills of artisans

8. Sociobiology is the best-known variant of essentialism and _____________ is its leading exponent.

a. Charles Darwin

b. Mark Granovetter

c. E. O. Wilson

d. Max Weber

9. By the age of about _____________, interaction with peers becomes an important factor in reinforcing gender-typed attitudes and behaviours.

a. 4

b. 8

c. 10

d. 14

10. Homosexuality was considered a serious psychiatric disorder from the early 1880s until _____________, when it was finally dropped from the *Diagnostic and Statistical Manual of Mental Disorders*, the standard diagnostic tool used by North American psychiatrists.

a. 1965

b. 1974

c. 1980

d. 1985

11. Sociologist Reginald Bibby predicts that by about 2010, only a durable core of some 15 percent of the population will continue to be opposed to _____________.

a. cohabitation

b. premarital sex

c. homosexuality

d. transsexuality

12. Social constructionists stress three main sociological changes that led to the development of gender inequality. Which of the following is *not* one of them?

    a. long-distance warfare and conquest

    b. the suffragette movement

    c. plough agriculture

    d. the separation of public and private spheres

13. A 2005 survey of 317,000 people provides recent information on sexual behaviour in 41 countries and concludes that ____________ have sexual intercourse most often—138 times per year, on average.

    a. Greeks

    b. Canadians

    c. Spaniards

    d. Americans

14. Which of the following is *not* a critique of the essentialist theory?

    a. Essentialists ignore the historical and cultural variability of gender and sexuality.

    b. The research evidence employed by essentialists is often flawed.

    c. Essentialists fail to consider the demographic shifts that occur over time.

    d. Essentialists ignore the fact that gender differences are declining rapidly.

15. Even today, many people assume that individuals should desire only members of the opposite sex. Sociologists call this assumption ____________.

    a. discrimination

    b. compulsory heterosexuality

    c. homophobia

    d. predetermined gender assumptions

Critical Thinking

1. The story of David Reimer is an incredible journey of courage and determination that puts forth a number of questions concerning the notions of sex and sexuality. In your opinion, should men and women that suffer from gender ambiguity be assisted in their struggle to find the right fit? As a society, how much assistance should we offer? For example, should the costs of counselling and surgery be covered by a provincial health insurance plan? Why or why not?

2. Gay marriages (and divorces) are now a common occurrence in Canada, but just recently the first lesbian couple at the Edmonton Prison for Women got married.

Although they currently reside in separate cells, they can apply to become cellmates and for all intents and purposes, live as what we would think in a heterosexual world as "man and wife." Are these inmates getting advantages that other inmates aren't? What are the implications if one of them were to be transferred?

3. Surveys are a popular method that provide evidence of a wide variation in attitudes toward sex or sexuality and, consequently, help to dispel the myths that surround these concepts. Since much of the information gathered by surveys relies on self-reporting, how reliable and valid is the information that we get from them? How honest are people likely to be when reporting information on such an intimate part of their lives? How reliable and valid is the data obtained from these sources?
4. According to the chapter, in most schools teachers still tend to assume that boys will do better in the sciences and mathematics and girls will excel in languages. It also says that parents reinforce these expectations at home. Do you agree or disagree with these statements? Were you "guided" into a gender-appropriate stream when you were in high school? How can you stop the perpetuation of these gender-biased stereotypes?

Time to Gear Down!

At the conclusion of this chapter, you should be able to discuss or write about the following, without having to rely on the textbook:

1. Sex refers to biological differences between males and females, while gender refers to the attitudes, beliefs, and behaviours that we commonly associate with each sex.
2. Although it's popular to trace the origins of masculine and feminine gender roles to biological differences, there are many ways in which gender is socially constructed.
3. Three major historical changes have led to the development of gender inequality: long-distance warfare and conquest, plough agriculture, and the separation of public and private spheres during early industrialization.
4. Conscious sexual learning begins around adolescence and occurs within the context of firmly established gender identities.
5. Although we receive little formal socialization regarding sexuality, sexual relationships tend to be male-dominated as a result of the character of gender socialization and men's continuing dominant position in society.
6. The social construction of gender and sexual scripts has defined standards of beauty that are nearly impossible for most women to achieve. This contributes to widespread anxiety about body image and, in some cases, leads to eating disorders.

7. Gender inequality and a social context that justifies and eroticizes sexual aggression in men contribute to the widespread problem of male sexual aggression.
8. The mass media reflect and reinforce the relationship between heterosexuality and male domination.
9. Social constructionism encourages sexual pluralism, which assesses the validity of sexual activities in terms of the meanings of the acts to the participants.

Web Links

From Sex to Humanity: How to Be Human—A Guide in Two Parts (Part 1)

http://www.firstscience.com/SITE/ARTICLES/human.asp

This website contains an article written by Peter Moore and provides an unbiased view on sexuality and the role that appearance plays in defining it. The article itself is a bit dated, especially because David Reimer has since committed suicide, but I really like how he explains sexuality.

Open Directory—Society: Sexuality

http://dmoz.org/Society/Sexuality

Included is an extensive listing of concepts and definitions that deal with sex and gender. It includes some pretty racy material and a host of online self tests, so I'm warning you now that some of this material might be offensive!

An Evolutionary Hypothesis for Eating Disorders

http://cogprints.ecs.soton.ac.uk/archive/00000800/00/eatdis~1.htm

This site offers an excellent sociological study of eating disorders.

Transgender at Work

http://www.tgender.net/taw

I've included this website as a real "eye opener" but also because I think it's an excellent example of bringing to light the reality that our world is truly changing—and our ideas and social policies surrounding sex and sexuality must also change.

Amnesty International Lesbian Gay Bisexual and Transgender Network

http://www.ai-lgbt.org/eng.htm

This site explains the extraordinary circumstances that these people suffer in countries all over the world. We take much for granted in Canada, and when you read some of these stories you'll be amazed to know that such treatment still exists in this world.

Critical "Linking" Question:

This one is a total no-brainer! Go to the Transgender at Work website and spend some time there looking at the Human Resource policies. Pay specific attention to the pages that explain how transgendered individuals should "reenter" the workforce after their transformations. If you were an employer, would you hire a transgendered person? No lying! What are some of the issues that you might face if one of your existing employees decided to become a "trannie"?

Solutions

True or False?

1. F
2. T
3. F
4. T
5. T
6. F
7. T
8. F
9. T
10. T

Multiple Choice

1. D
2. D
3. C
4. D
5. B
6. A
7. A
8. C
9. D
10. B
11. B
12. B
13. A
14. C
15. B

Chapter 5
The Mass Media
Chapter Introduction

Tell the truth—how many television sets are in your home? Just this past term, a student told me that he had 12 televisions in his house! I was—and still am—astounded by that! Without a doubt, the media, in all its forms, have a staggering effect on our lives, regardless of how often we watch television or read the daily news. It is virtually impossible to live anywhere without having access to some form of media—television, radio, newspapers, magazines, the Internet and billboards are all types of media that keep us informed on what's happening and what's not happening—sometimes when we really don't care about either. Though I could easily live without a television, I couldn't live without a computer and I would give up food before I gave up books. If you had to relinquish one or the other for a month, which would it be? How much would it change your life? Are you easily duped into buying items that you really don't need because you like the product's advertising campaign? Though I try not to, I sometimes fall prey to impulsive shopping and buy things based solely on the commercials that advertise them. Do magazines filled with pictures of beautiful women and hunk-a-licious men make you feel less than standard? In this image-based society, where you rarely get a chance to see anyone that look like an actual human being in the media, how do you form your self-perceptions and to what extent do the beautiful models—both male and female—affect your body image? I can only hope that you realize that absolutely no one in those magazines actually looks that way! The wonders of airbrushing and graphic design are amazing these days—those models don't even have skin pores!

This chapter explores the different forms of media that affect our lives and offers a variety of theoretical viewpoints that help us understand the media as an important influence in our society. By addressing four main ideas that surround the media, the author gives some very clear ideas concerning political economy, representation and ideology, media effects and audiences, and the Internet.

Jump Start Your Brain!

In this chapter you will learn that:

1. The mass media may be examined in terms of their economic and political organization, their representation of ideas, and their effects on people.
2. Newspapers are local monopolies with high levels of ownership concentration; their dependency on advertising has an impact on news content because it limits the survival prospects of newspapers with radical views.
3. English-speaking Canadians watch mostly American television programming, especially prime-time drama and comedies. Advertising dependency, audience

preferences, and the lower cost of American programs discourage the production of Canadian television drama. Although some analysts view this as a "sellout" of Canadian culture, others regard it as an effect of globalization and argue that it does not undermine the institutional structure of society.

4. Some analysts claim that news coverage has a left-liberal political bias, while other analysts argue that news coverage is ideologically conservative.
5. Many analysts believe that a causal link exists between media violence and violent behaviour, but studies that demonstrate this alleged connection have been criticized on the grounds of flawed methodology. Even those who believe that media violence causes aggressive behaviour disagree about how it does so.
6. Studies on the viewing habits of TV audiences indicate that women and men not only prefer different types of programs but also watch TV in different ways. Men tend to use TV in a more planned way than women do and watch more intently, whereas women are more likely to use TV as a focus for social interaction.
7. Access to and use of the Internet reflects broader patterns of social inequality, though this appears to be in decline as the technology becomes less costly and more widely available. Use of the Internet for social interaction creates virtual communities that act as sources of identity and social support. New communication technologies also provide opportunitics for radical forms of creativity and the development of alternative social arrangements.

Quiz Questions

True or False?

1. Despite the decline in television viewing, research indicates that we spend more time interacting with the media than doing anything else, including working. True or False
2. Space-biased media are modes of communication that endure over time but are not very mobile across space, such as writing on stone or clay tablets. True or False
3. Political economy focuses on the ownership and control of economic resources, and on the effect of technology and economic power on cultural values, social structure and political decision making. True or False
4. The critical perspective has three variants. One that emphasizes the relationship between media and inequality, one that emphasizes the relationship between media and social conflict and a third that focuses on the relationship between the media and the individual. True or False
5. Desensitization refers to the possibility that continued exposure to violent imagery may weaken the mechanisms of self-control that an individual acquires through socialization and that discourage the use of violence. True or False

6. The world's dominant media corporations are primarily American-based. Only two of the top six media giants are headquartered outside of the United States. True or False
7. Virtual communities resemble real communities inasmuch as they entail not only shared values and a sense of belonging but also a common project or purpose. True or False
8. As social structure becomes increasingly differentiated, social roles, identities, and experiences become increasingly fragmented and dissociated from one another. Media specialists believe that fragmentation also allows groups and individuals greater scope to play with and reconstruct their identities and understandings. True or False
9. Horizontal integration refers to the controlling of resources and assets at the different stages of production, such as ownership of a major league sports team along with the stations and cable channels over which the games are televised. True or False
10. Early research and commentary on the Internet tended to polarize between optimists who saw CMC positively (as a vehicle for greater democracy, globalism and identity experimentation) and those critics who focus on such negative effects as disengagement from "real" social relations and the spread of offensive images and ideas like pornography and racism. True or False

Multiple Choice

1. Access to and use of the Internet reflects broader patterns of ____________, though these may decline as the use of technology becomes less costly.

   a. educational attainment

   b. gender inequality

   c. social inequality

   d. regional dependence

2. The most commonly used form of CMC among Canadians is ____________.

   a. e-mail

   b. general browsing

   c. formal education/training

   d. purchasing goods and services

3. When the media focuses on certain issues while playing down or ignoring others, the process is called ____________.

   a. framing

   b. selective representation

   c. biased reporting

   d. priming

4. According to the ____________ perspective, institutions such as the media and processes such as socialization and social control cannot be understood from the viewpoint of society as a whole, but only from that of unequal and conflicting groups and classes.

   a. structural functionalist

   b. critical

   c. symbolic interaction

   d. essentialist

5. In 2003, the average Canadian spent just less than 20 hours a week listening to radio and ____________ hours a week watching television.

   a. 8.5

   b. 17.5

   c. 21.7

   d. 30

6. In the realm of MUDs, newsgroups and chat rooms, lurking can act as a form of __________ by allowing newbies to familiarize themselves with the group's practices and language codes before becoming active posters.

   a. primary socialization

   b. anticipatory socialization

   c. secondary socialization

   d. resocialization

7. News sources that do not have an organizational or group affiliation but are usually eyewitnesses or victims of newsworthy events and issues are called ____________.

   a. ordinary news sources

   b. alternative news sources

   c. unreliable news sources

   d. independent news sources

8. The ___________ perspective sees multimedia chain ownership, local monopolies, and advertising dependency as resulting in the homogenization of news coverage and the decline of diversity in the news topics and view points.

   a. technical

   b. ideological

   c. critical

   d. hegemonic

9. News ____________ tends to focus on the activities of certain social actors rather than others, so it also tends to rely on and privilege certain sources of information over others.

   a. streaming

   b. marking

   c. censuring

   d. framing

10. Edward Herman and Noam Chomsky argue that the media serve the interests of the political and economic elites by ____________ information to reduce or eliminate radical or subversive views.

    a. framing

    b. filtering

    c. suppressing

    d. ignoring

11. The issue of Canadian content pertains largely to entertainment programming, particularly English Canadian drama. Roughly ____________ of all English-language viewing in Canada is devoted to American television, and this number climbs to ____________ during prime time.

    a. 70%; 90%

    b. 40%; 60%

    c. 50%; 70%

    d. 60%; 80%

12. Which of the following is *not* considered to be one of the three major criteria used to determine newsworthiness?

    a. potential worth

    b. immediacy

    c. personalization

    d. extraordinariness

13. ____________ refers to a situation in which one society's media exert an overwhelming and unilateral influence over another society's culture.

    a. Cultural domination

    b. Cultural hinterland

    c. Cultural hierarchy

    d. Cultural imperialism

14. Conservatives argue that the news media have a "left-liberal" bias that runs counter to the views and interests of mainstream of society. According to these

conservative writers, bias operates in three ways. Which of the following is *not* one of them:

a. an anti-corporate bias

b. give greater or more favourable attention to the views of interest groups and constituencies that share their personal political views

c. hire reporters and journalists that espouse their own beliefs and ideologies

d. concentrate on negative events, issues and news angles

15. Initial access to the Internet is typically marked by a strong gender bias and is particularly evident in Canada and _____________, where women now (slightly) outnumber men online.

a. Japan

b. Germany

c. France

d. the United States

Critical Thinking

1. Ownership and control of the media arc gcnerally becoming more concentrated into a smaller number of larger corporate hands. Are you aware of the relationship that exists between the media and big business? As a consumer of a variety of media forms, how does this relationship affect you? For example, how does network ownership affect the news? The textbook example of CanWest Global and the *Ottawa Citizen* should offer some insight on this question.

2. Researchers have found that the introduction of TV has led to the displacement of other activities that have a social element attached to them. Has the availability of stay-at-home shopping, banking, working, and entertainment diminished the time that you would normally spend doing these things in person? Can you imagine a time when you won't ever have to leave your home? Would you find other ways to socialize or would you be content just staying indoors?

3. Television is a heterogeneous medium and its imagery is usually open to some interpretive variation. The degree of variation depends partly on the modality of the image—that is, the extent to which it approximates real life. Has the increase in "reality TV" far surpassed that which is real? Give this some serious thought now—how much of what is shown on television is worth the time it takes to watch it?

4. Those who accept the majority view claim that there is consistent evidence for a link between television and real-life violence or aggressiveness, though the strength of the evidence varies according to the methodology. If you had the chance to research the relationship between television and aggression in children, how would you define and measure the concept of violence? Would your definitions and measurements differ if your sample comprised adults instead of children?

Time to Gear Down!

At the conclusion of this chapter, you should be able to discuss or write about the following, without having to rely on the textbook:

1. Canadian newspapers have high levels of ownership concentration, are usually part of larger multimedia chains, and function largely as local monopolies. The factor that has the greatest impact on news content, however, is dependency on advertising. In fact, the survival prospects of papers with an alternative or more politically radical view of the news are limited because of it.
2. In the case of television, advertising dependency, the high costs of production, and audience preferences mean that much of the programming Canadians watch is foreign, primarily American. This is especially so for drama and sitcoms, and particularly among anglophones. Nationalists view the situation as a sellout of Canadian culture, whereas postmodernists see it as part of the general effect of globalization and do not believe that it undermines the institutional structure of Canadian society.
3. Conservatives claim that news coverage has a left-liberal political bias that is unrepresentative of society's mainstream. Critical theorists argue that news coverage is ideologically conservative, in that it defines reality from the perspective of dominant ideology and views events and issues through the lens of social control.
4. At the same time, some critical theorists argue that the media are not completely closed to alternative voices and viewpoints, and that these can challenge, to some extent, the hegemony of dominant social groups and interests.
5. The majority of observers believe that a causal link exists between television and violent behaviour, but the studies that support this view have been criticized on the grounds of flawed methodology. Moreover, the majority view is split on the issue of how television causes aggression. Some argue that watching TV counteracts the effects of socialization by weakening self-control; others believe that it, in fact, socializes children in the use of violence.
6. Studies of audiences indicate that TV viewing and the responses to it vary according to gender. Men and women tend to prefer different types of programming. Men are more likely to watch attentively and privately, whereas women watch in a more interactive, social way. Women also tend to be more open about their TV viewing and use TV as a topic of casual social interaction and conversation.
7. Although early views about the development of computer-mediated communication and the Internet tended to polarize between optimists and pessimists, recent research has yielded a more complex, balanced view.
8. Inequalities of access (especially globally) and differences of use persist; however, the former show signs of declining as the technology becomes less costly. Research on the impact of the Internet shows that virtual communities develop normative structures like real communities, and function as sources of

identity and social solidarity. For the most part, however, life online supplements and complements real-world social interaction and involvement rather than replacing it. This applies as much to social clubs and fan groups as it does to activists.

Web Links

Accuracy in Media—For Fairness, Balance and Accuracy in News Reporting

http://www.aim.org

Accuracy in Media is a non-profit, grassroots citizens watchdog of the news media that critiques botched and bungled news stories and sets the record straight on important issues that have received slanted coverage. It also contains some very-well-written commentaries by guest columnists on a variety of interesting issues. Definitely worth checking out!

Media Watch: Media Literacy Through Education & Action

http://www.mediawatch.com

Media Watch, which began in 1984, distributes educational videos, media literacy information, and newsletters to help create more informed consumers of the mass media. Totally take some time to really check out what is going on in this website—the picture gallery is amazing—it has some of the most graphic advertising images you'll ever see!

CRTC Broadcasting—Cultural Diversity Policy

http://www.crtc.gc.ca/eng/INFO_SHT/b308.htm

This link will bring you to the cultural diversity policy of the Canadian Radio and Television Broadcasting Corporation, and I suspect you'll be surprised to find out the regulations surrounding air play. Click on the "Home Page" link for further information.

Nielsen Ratings

http://www.nielsenmedia.com/nc/portal/site/Public

Okay—this has to be one of the coolest sites I have ever been on! Not only does it show you the most talked about ratings in television, it actually explains the methodology behind the collection of those ratings. An extremely professional site that's just jammed with information for prospective employees, students, and just plain people checking out the ratings!

Media Awareness Network

http://www.media-awareness.ca/english/index.cfm

Resources and support for everyone interested in media and information literacy for young people.

Critical "Linking" Question

As if you didn't see this one coming! Go to the Nielsen Ratings website and click on the Inside the Ratings link. Take a good look at how they collect their data and see if you can find any flaws with their reports. Is their sample representative? What about the monitoring processes?

Solutions

True or False?

1. T
2. F
3. T
4. F
5. F
6. T
7. T
8. F
9. F
10. T

Multiple Choice

1. C
2. A
3. D
4. B
5. C
6. B
7. A
8. C
9. D
10. B
11. C
12. A
13. D
14. C
15. D

Chapter 6
Social Stratification

Chapter Introduction

I often ask my students if they'd rather win $5 million dollars—I even make it tax-free—or the Nobel Peace Prize. Year after year, the answers are inevitably the same—98 percent of my students will choose the money. But this year a funny thing happened. One eager student quickly raised his hand and said "I'd take the $5 million dollars and then I could *buy* my own Nobel Peace Prize!" What do you think of that? Well, I'll tell you, it didn't take me long to correct this bright young man and tell him that his words would soon come back to haunt him. He quickly learned that money can't buy everything and that the Nobel Peace Prize, though it comes with a $1 million dollar prize, is about so much more than money. It's all about the prestige and honour that goes with it—just like the concept of class is more than about how much money you have. If you won $22 million in the lottery, how would it affect your social class? Though you'd likely move to a nicer house and drive a better car, do you think that your winnings would gain you entry into the world of the wealthy elite? Would it give you power and prestige? Can you imagine getting an invitation to dine with Bill Gates or Stephen Harper? I can't imagine that you would. Sociologists study social class for a variety of reasons and they almost always come up with a variety of interesting results, but one thing is usually very certain—there is a distinct difference between social classes in Canada and this difference is made up of more than just money.

In this chapter, you'll learn about some of the ways in which sociologists study stratification and the theories that attempt to explain why social inequalities exist in society. Additionally, you'll take a look at occupational and class structures, as well as material inequality in Canada. It really is amazing to examine the disparities that exist between classes. Lastly, you'll investigate whether or not inequality has been increasing in the recent past.

Jump Start Your Brain!

In this chapter you will learn that:

1. Persistent patterns of social inequality are based on statuses assigned to individuals at birth and on how well individuals perform certain roles. Societies vary in the degree to which mobility up and down the stratification system occurs.
2. Explanations of the origins and impact of social stratification include the theory of Karl Marx, which emphasizes the exploitation of the working class by owners of land and industry as the main source of inequality and change; the theory of Max Weber, which emphasizes the power that derives from property ownership, prestige and politics; structural-functionalist theory, which holds that stratification

is both inevitable and necessary; and several revisions of Marx's and Weber's ideas that render them more relevant to today's society.

3. Although there has been considerable opportunity for upward occupational mobility in Canada, wealth and property are concentrated in relatively few hands and one in seven Canadians lives in poverty. Because of labour-market changes, income and wealth inequality have been increasing. Thus, the stratification structure of the future will probably not resemble the pattern that emerged in the affluent middle of the twentieth century.
4. A person's position in society's stratification system has important consequences, both for lifestyle and for the quality of life. Those who are situated higher in the economic hierarchy tend to live better and live longer.

Quiz Questions

True or False?

1. An ascribed status can be a function of race, gender, age, and other factors that are not chosen or earned, and that cannot be changed (a few people do choose their gender status, they are rare exceptions). True or False
2. Marx identified two major classes—the capitalist class or proletariat, which owned the means of production—and the bourgeoisie, or working class, which exchanged its labour for wages. True or False
3. The economic hierarchy is not completely closed, but it is relatively stable and permanent, and is composed of individuals with similar amounts of control over material resources. True or False
4. Marx reasoned that the value of a product sold was directly proportional to the average amount of labour needed to produce it. Thus, for example, an elegant piece of furniture was more valuable than its component pieces mainly because of the labour invested in it by the worker(s) who built it. True or False
5. Weber proposed that structures of social stratification could be better understood by looking at class, status, and party. True or False
6. Since Marx lived to see the emergence of white-collar workers, the growth of large private- and public-sector bureaucracies, and the growing power of trade unions, he was able to write about these alternative sources of power in a stratified capitalist society. True or False
7. According to Davis and Moore (1945), social inequality is both inevitable and functionally necessary for society. True or False
8. According to Marx, conflict in modern society was diffused across many different sets of competing groups and was much less likely to lead to significant social unrest. True or False
9. As the Economic Council of Canada (1992) indicated, poverty is not a static status. A sizeable number of Canadians move in and out of poverty each year. True or False

10. The gender wage gap becomes smaller when looking at older, university-educated workers. True or False

Multiple Choice

1. In a ____________ everyone would have equal chances to compete for higher-status positions and, presumably, those most capable would be awarded the highest rank.

   a. socialist system

   b. meritocracy

   c. democracy

   d. capitalist system

2. __________ updated the original Marxist model so it could be applied to the twentieth century.

   a. Erik Olin Wright

   b. Lenski

   c. Davis and Moore

   d. Dahrendorf

3. ________________ refers to a position of an individual or family within an economic hierarchy, along with others who have roughly the same amount of control over access to economic or material resources.

   a. Status

   b. Prestige

   c. Societal level

   d. Class

4. On average, women earn about _____________ percent of what men earn.

   a. 50

   b. 60

   c. 70

   d. 80

5. According to the writings of Marx, which of the following economic forces did *not* contribute to the creation of a class-based society?

   a. emergence of large mechanized factory-based systems of production

   b. movement of women from the home to the workplace

   c. rapid growth of cities as rural peasants left the land

   d. extreme material inequality caused by huge profits for factory owners and merchants and the poverty of labourers

6. Weber shared with Marx a belief that economic inequalities were central to the social stratification system and that the ownership of ___________ was a primary determinant of power, or the ability to impose your wishes on others, to get them to do what you want them to do.

   a. property

   b. wealth

   c. workers

   d. power

7. The most prominent occupational shift over the course of the twentieth century is the decline in __________________ occupations.

   a. agricultural

   b. mining

   c. self-employed

   d. factory

8. Whereas capitalists might be conscious of their group interests, wage-labourers needed to become aware of their common enemy. They needed to be transformed from a "class in itself" to a "class for itself." Thus, __________ was an important social-psychological component of Marx's theory of social inequality and social change.

   a. class consciousness

   b. false consciousness

   c. total awareness

   d. solidarity

9. __________________ refers to the occupational mobility that occurs within a society when better-qualified individuals move upward to replace those who are less qualified and who must consequently move downward.

   a. Circulatory mobility

   b. Vertical mobility

   c. Horizontal mobility

   d. Intergenerational mobility

10. Which of the following is *not* a factor in determining socioeconomic status?

    a. income

    b. education

    c. occupation

    d. gender

11. The wealthiest 10 percent of families, with an average net worth of $1,320,900, held almost _____________ of all wealth in Canada.

    a. one-quarter

    b. one-half

    c. one-third

    d. three-quarters

12. Yakabuski (2001) demonstrated that in 2000 _____________ of the companies listed on the Toronto Stock Exchange 300 Index were each controlled by a single owner holding more than 50 percent of the voting shares.

    a. two-thirds

    b. one-third

    c. one-quarter

    d. one-half

13. Which of the following is *not* a characteristic of Weber's typology of inequality?

    a. employment

    b. class

    c. status

    d. party

14. Marx argued that the value of goods produced by wage-labourers far exceeded the amount needed to pay their wages and the cost of raw materials, technology, and other components of the means of production. Marx referred to this excess as ____________.

    a. competitive advantage

    b. surplus labour

    c. surplus exchange

    d. surplus value

15. Dramatic changes in the status of various groups have occurred in this country over time. Although the practice was not nearly as widespread in Canada as in the United States, slaves were bought and sold in Canada until the __________. By then, slavery had disappeared in Canada and we now have laws against discrimination on the basis of race.

    a. 1600s

    b. 1700s

    c. 1800s

    d. 1900s

Critical Thinking

1. "Are social classes dying?" I'm wondering whether or not you think that class distinctions are very prominent today and if so, how you can tell the difference between them. Imagine describing Canadian society to someone from another country without referring to some features of stratification—or the class structure—in Canada. Write a letter to a friend from another country who is asking whether or not poverty exists in Canada.
2. A teenager from a wealthy family graduates from an excellent high school in an affluent neighbourhood, completes a degree or two in a prestigious and costly university, and then begins a career in a high-status, well-paying profession. Is this an example of someone achieving a deserved high-status position, or did the advantages of birth play some part in this success story? What do you think? Although some students may have monetary advantages over others, do they have any other kind of advantage over other students? If so, what are these advantages?
3. Although some differences in pay might be justified to reimburse those who spend more years in school preparing for an occupation, are the huge income inequalities that we see in our society necessary? Are movie stars, professional athletes, and chief executive officers with million-dollar-plus annual incomes more important to society than nurses, daycare workers, prison guards, and most other lower-paid workers? If you have plans to become a professional—a doctor for example—would you consider becoming a missionary or would you expect to earn a salary that would offset the cost of your investment?
4. King or Queen for a Day!!! Now that you've read all these theories on why the different classes exist in society—and if you're like me, you can't quite choose one that explains it just the way you like—throw them down the drain for a second and pretend that you're King or Queen for a Day and that you're in charge of running our country. What strategies would you put in place to decrease the income gaps that exist between rich and poor? Would you put a cap on salaries? Could you take away our social welfare system—would you?

Time to Gear Down!

At the conclusion of this chapter, you should be able to discuss or write about the following, without having to rely on the textbook:

1. Persistent patterns of social inequality within a society are referred to as a structure of social stratification. Some social hierarchies within a society are based on ascribed statuses, such as gender, race, or age, which are typically assigned to an individual at birth. Other social hierarchies are based on achieved statuses, which index how well an individual has performed within some role. A society in which considerable social mobility between statuses is possible is said to have an open stratification system.
2. Social theorists have proposed a variety of different explanations of the origins and effects of social-stratification systems. In his class-based theory of social

stratification, Karl Marx emphasized the exploitation of the working class by the owners of the means of production and the capacity of class conflict to generate social change. Max Weber also put considerable emphasis on the power that resides in ownership of property but argued that other hierarchies, those of prestige and political power, are influential as well.

3. The structural-functionalist theory of social stratification suggests that inequality is both inevitable and functionally necessary for a society, ensuring that the most qualified individuals are selected to fill the most important (and most rewarding) roles. Power differences are downplayed in this theory, as is conflict between different social classes. A number of more recent theories of social stratification, including those put forward by Gerhard Lenski, Frank Parkin, and Erik Olin Wright, have placed much more emphasis on power and conflict. Although Wright developed a class-based theory of stratification that adapts many of Marx's ideas to contemporary circumstances, Parkin's approach follows in the footsteps of Weber.

4. Examination of occupational shifts in Canada over the past century reveals some of the changing features of Canada's stratification system. However, we can get only partial glimpses of the class structure of our society. Studies of occupational mobility reveal that Canada is a relatively open society. Even so, considerable evidence suggests that class-based advantages are passed from one generation to the next.

5. A detailed analysis of material inequality in Canada reveals that ownership of wealth and property is highly concentrated and that income inequality is relatively high. By using a relative measure of poverty, we observe that about one in seven Canadians is living below the "poverty line." Considerable evidence shows that the poor and others near the bottom of the social hierarchies in our society enjoy fewer life-chances than do the well-off. For example, they are less likely to do well in school and to continue on to higher education, they are less healthy and have a shorter life expectancy, and they do not fare as well when dealing with the criminal justice system. But because of their more limited access to social and material resources, the poor have seldom become an active force for social change.

6. Some theories of social stratification developed in the middle of the twentieth century suggested that material inequality was declining as the North American economy expanded. However, the period of rapid economic growth and relative affluence that characterized the middle decades of the century appears to have ended. And, as unemployment rates rise, as part-time and temporary work becomes more common, and as governments cut back on social-assistance programs, evidence accumulates that material inequality is also slowly increasing in Canada.

Web Links

Forbes—20 Richest Women in Entertainment

http://www.forbes.com/digitalentertainment/2007/01/17/richest-women-entertainment-tech-media-cz_lg_richwomen07_0118womenstars_lander.html

It's likely not a surprise that Oprah Winfrey tops this list. Take a surf through some of the other salaries and some of the other categories for a very real look at the outrageous amount of money that Hollywood stars rake in.

The Caste System

http://www.friesian.com/caste.htm

This site offers a very interesting—and accurate—description of the caste system of hierarchy in India. It's extremely cool to read how different a country can differ so much from our own!

Raising the Roof

http://www.raisingtheroof.org/lrn-hh-index.cfm

I've brought you to a page of this site that explains a few dimensions of "hidden homelessness" and mentions the working poor—like in your text. I'd like you to take some time though, and really check this site out—maybe there's something that you can do to help!

Canadian Council on Social Development

http://www.ccsd.ca/home.htm

The Canadian Council on Social Development (CCSD) is a non-governmental, not-for-profit organization that was founded in 1920. Their mission is to develop and promote progressive social policies inspired by social justice, equality, and the empowerment of individuals and communities.

Barratt's Measure

http://wbarratt.indstate.edu/socialclass/Barratt_Simplifed_Measure_of_Social_Status.pdf

OK—this read might be a little tricky, but see if you can tell whether or not this would be an accurate measure of a person's social status. Give it your best shot!

Critical "Linking" Question

This one is a contentious question and one that I'm sure will ruffle more than a few feathers. Check out Oprah's salary on the Forbes website. Though I don't watch her television show (I don't watch television in general), I do know that she often gives out expensive gifts to the audience—and on one show, gave every member of the audience a new car! Do you think that this money could/should have been better spent? How do you feel about her "on air" generosity?

Solutions

True or False?

1. T
2. F
3. T
4. T
5. T
6. F
7. T
8. F
9. T
10. F

Multiple Choice

1. B
2. A
3. D
4. C
5. B
6. A
7. A
8. A
9. A
10. D
11. B
12. C
13. A
14. D
15. C

Chapter 7
Gender Inequality: Economic and Political Aspects

Chapter Introduction

Do you think that the career to which you're aspiring will be affected at all by your gender? Are you aware of the inequalities that exist between men and women, especially when it comes to paid and unpaid work? When was the last time you thanked your mom (or in a few cases, your dad) for doing your laundry or cleaning up the kitchen? I can remember the first time I ever heard a male telephone operator. I was really caught off guard because it was just so unheard of at the time. Now, because of all the telemarketers, I'm used to hearing men in that role much more often. I had the same reaction when I first saw a man behind a checkout counter in the grocery store. It's not like men weren't qualified or entitled to do those jobs, it was just that they had traditionally been performed by women and I wasn't accustomed to seeing men in these roles. It's strange that I would have had that reaction, because, a long time ago, I trained to become a machinist and, for a while, was one of the few female machinists in Ontario. I think it was then that I first learned that the world of work was very much different for women than it was for men—especially if you were a woman in a traditionally male-dominated job. The tricks that my fellow male machinists used to play on me were merciless and though I quit that job, years later I worked two summers at GM to make money for tuition and once again, found myself in the same position. Though it was hard to believe, things hadn't changed much and the crazy antics between the guys and the girls in the shop still continued. I'll never forget the time I went into work only to find that my male coworkers had spray-painted my work boots pink! Yet even harder to believe is that gender inequality still exists in Canada and it doesn't seem as though it will ever even out, let alone disappear.

This chapter begins with a definition of gender inequality and then explores the major dimensions of it in Canadian society, with specific reference to the home, the labour force, and politics. You'll learn all about gender inequality as it exists in the economic and political aspects of society as well. Much of what you read will probably astound you, and I hope that you get a good handle on the inequalities you may have to face one day. The chapter ends on a bright note, though, as the author discusses the actions, policies, and legislation that could reduce gender inequality in the future.

Jump Start Your Brain!

In this chapter you will learn that:

1. A major source of inequality between women and men is the greater exclusion of women from public economic and political activity. The persistent tendency of

many women to be relegated to domestic affairs leads to less income, prestige, and power.

2. Although women have entered politics and the paid labour force in increasing numbers over the past century, they still tend to be chiefly responsible for meal preparation, cleaning, laundry, and child-care.
3. In the paid labour force, women and men still tend to be segregated in different kinds of jobs; "women's jobs" typically pay less, carry less prestige, offer less security, and bestow less authority.
4. Although women make up more than half the Canadian population, only one in five members of Parliament is a woman.
5. In recent years, employment equity policies and policies requiring equal pay for work of equal value have been implemented to help lessen gender inequalities, but much research and political action are still required before gender equality is achieved.

Quiz Questions

True or False?

1. Women are usually employed in lower-skilled jobs. Since higher-skilled jobs usually pay more and offer more security, employment in lower-skilled jobs implies economic and quality-of-work inequalities between men and women. True or False
2. Nepotism is the average evaluation of occupational activities and positions that are arranged in a hierarchy. It reflects the degree of respect, honour, or deference generally accorded to a person occupying a given position. True or False
3. In the 90 years since most women gained the vote federally, only three women have been elected as party leaders, all within the last two decades. True or False
4. When women face invisible barriers in penetrating the highest levels of organizations where power is concentrated and exercised, it is referred to as the glass ceiling effect. True or False
5. Oversimplified beliefs about how men and women, by virtue of their physical sex, possess different personality traits and as a result, may behave differently and experience the world in different ways are called gender labels. True or False
6. Thelma McCormack (1975) suggests that gender differences in voting interests may be due to the fact that women and men operate in different political cultures (which have been moulded by gender differences in political socialization) and have different opportunities to participate in politics. True or False
7. In husband-wife families, wives' earnings not only increase family incomes but also help to keep many families out of poverty, particularly when the husbands earn little or nothing. True or False

8. Between the 1970s and early 1990s, the labour force participation rate of women more than doubled, and the number of elected women MPs quadrupled. True or False

9. Since 1980, women have made up around 85 percent of the part-time labour force. True or False

10. Gender discrimination is the process whereby employers make decisions about whether to hire and how much to pay any given woman on the basis of the employers' perceptions of the average characteristics of all women. True or False

Multiple Choice

1. In 1996, a Statistics Canada economist estimated that unpaid domestic work, if done in the market for wages, would be worth about ____________ of the gross national product (the total value of market-produced goods and services produced in the country over a year).

   a. three-quarters

   b. one-quarter

   c. one-half

   d. one-third

2. The growing labour-force participation of women has changed the composition of the Canadian labour force. At the beginning of the twentieth century, only ____________ percent of women were economically active in the paid labour force compared with 78 percent of men.

   a. fourteen

   b. twenty

   c. ten

   d. eighteen

3. The fact that gender is largely learned and that its content is continually renewed and altered through social interaction has three implications. Which of the following is *not* one of them?

   a. Gender identification can often be manipulated to match a person's sex (for example, transgendered individuals).

   b. Gender identities and behaviours are not stable and fixed.

   c. Gender identities and gender-specific behaviours are not always congruent with the sex assigned at birth.

   d. Gender identities and behaviours are not binary and polar opposites.

4. Which of the following did not influence changes in the labour-force participation rate of Canadian women?

   a. women choosing to marry at an older age

b. increased demand for workers in service jobs

c. decreases in the number of children born

d. increased financial pressures on the family

5. Canada's fertility rates ____________ during the 1930s and early 1940s as a result of the Depression and World War II.

   a. increased substantially

   b. decreased substantially

   c. stayed the same

   d. were not measured

6. Observers have offered four sets of explanations for the pay gap between women and men. Which of the following is *not* one of them?

   a. biological differences, leading to job discrimination

   b. gender differences in the type of work performed

   c. societal devaluation of women's work

   d. gender differences in the characteristics that influence pay rates

7. Data from the 2001 census show that women are still more likely than men to do unpaid work involving home maintenance and child-care. For example, compared with married men, higher percentages of married women spend ____________ hours or more on housework and on child-care among those who have at least one child under age 15 in the household.

   a. 30

   b. 15

   c. 25

   d. 10

8. ____________ is the discrimination that occurs when negative decisions concerning the hiring or promotion of a given individual are made on the basis of the average characteristics of the group to which the individual belongs.

   a. Gender discrimination

   b. Statistical discrimination

   c. Nepotism

   d. Selective discrimination

9. The persistence of stereotypical thinking of feminine and masculine characteristics as polar opposites should sensitize us to two things. Which of the following does *not* sensitize us?

   a. In these polarized depictions, feminine traits are viewed as less desirable than masculine ones.

b. The idea of difference is apparently a powerful one, and hard to dispel even when it is contradicted by research.

c. Since the ability to control and influence—to use power—indicates the twin processes of domination and subordination, most sociologists describe the power relations between men and women as those of male domination and female subordination.

d. Living in a patriarchal society perpetuates the inequalities that exist between men and women.

10. Which of the following is *not* a characteristic of media descriptions of female politicians?

    a. They use feminine traits to manipulate their public image.

    b. They fail to recognize the prior political activities of female politicians, with the result that the women's histories of acquiring competency remain unknown.

    c. They suggest that female politicians are responsible for women's issues, when, in fact, gender interests may or may not be on the agenda of any politician, male or female.

    d. They use the term *feminism* or *feminist* to denote negative personal characteristics.

11. State intervention influences the magnitude of gender inequality and sustains or minimizes male relations of power over women in three areas. Which of the following is *not* one of them?

    a. reproduction

    b. family

    c. professional sports

    d. the labour force

12. _____________ refers to the body of thought on the cause and nature of women's disadvantages and subordinate position in society and to efforts to minimize or eliminate that subordination.

    a. Affirmative action

    b. Racial inequality

    c. Feminism

    d. Employment equity

13. _____________ is the result of the general belief that men are superior to women and may impose their will on them.

    a. Discrimination

    b. Sexual harassment

c. Inequality

d. Domination

14. ____________ are the behaviours that are expected of people occupying particular social positions.

a. Gender roles

b. Gender stereotypes

c. Social roles

d. Gender status

15. Multiracial feminism contributes to our understanding of gender inequality in three ways. Which of the following is *not* one of them?

a. It supports the ideological tenets of a matriarchal society.

b. It highlights differences among women in terms of gender inequality.

c. It points out that women of specific races and in certain class locations are in positions of power and domination over other groups of women.

d. It emphasizes that solutions to gender inequality vary according to the location of groups of women in the matrix of domination.

Critical Thinking

1. High wages, low wages—no wages? What do you think wages will be like when you graduate? Will the difference in wages between men and women still be as great? Will the glass ceiling be shattered? Box 7.1 on page 179 of your text shows the results of a study that indicates that barely a crack has been made in that ceiling, which seems to be made of bulletproof glass. What do you think, girls—will you be the first to shatter it?

2. When we have such rigorous requirements and extensive testing involved to gain entry into "skilled" professions—such as doctors, nurses or electricians—how can it be that our definitions of "skill" are socially constructed? According to Gaskell (1986), what we define as skill reflects other social hierarchies based on gender. Do you think that this explains why women are less likely than men to be employed in high-skilled jobs? Why or why not?

3. Even though the existence of "stay-at-home dads" is on the rise, women are still responsible for the majority of the housework that's done. What do you think would happen if the government devised a scheme to legitimize housework as paid labour? In your opinion, would the percentage of men who spend time on housework increase? Why or why not?

4. Are you planning to enter a gender-specific occupation? If you are then you're probably aware of the obstacles that you're going to have to overcome. Try listing some of what you think will be your biggest battles. Have you gathered your weapons? Have you developed your strategies to conquer them? What are your game plans?

Time to Gear Down!

At the conclusion of this chapter, you should be able to discuss or write about the following, without having to rely on the textbook:

1. Many sociologists view the segregation of women and men into the private and public spheres as a very important source of gender inequality. Exclusion from the economic and political arenas of Canadian life can mean disadvantages in access to income, economic well-being, prestige, and power. Restricted in the past to the domestic sphere, women have been economically disadvantaged and have had little or no opportunity to influence legislation dircctly. In addition, their unpaid work in the home has been considered low in prestige, or at least lower in value than their spouses' paid work.

2. During the twentieth century, women have entered the labour force and the political arena in ever-increasing numbers. Today, more than half of all women are in the labour force. Politically enfranchised, women have also entered the political arena, either as politicians or in connection with groups associated with the women's movement. Many of these changes have occurred—or, at least, have accelerated—since the time most of you were born. Between the 1970s and the early 1990s, the labour-force participation rate of women more than doubled, and the number of elected women MPs quadrupled.

3. If the glass is half full, it also is half empty. Although they do paid work, many women are still responsible for most of the meal preparation, cleaning, and laundry needs of their families. Women still tend to be considered the primary caregivers of children and of seniors. In short, women are more likely than men to work a double or triple "shift" every day.

4. In the labour force, women and men are occupationally segregated, with women concentrated in jobs stereotyped as "women's jobs." Women are more likely than men to be employed in jobs that are part-time or otherwise nonstandard. They earn less than men, on average, and their skills tend not to be fully recognized or fairly evaluated. To be sure, these issues affect women to varying degrees, depending on their birthplace, race, and ethnicity. Nevertheless, the overall picture is one of gender inequality in the labour force, with women disadvantaged relative to men.

5. There is still evidence of a gender gap in politics as well. Women represent more than half of Canada's adult population, but only 21 percent of federally elected legislators. This imbalance notwithstanding, substantial gains have been made in recent elections. And, as agents pressing for improvements in their own status, Canadian women have left a considerable legacy of influence and change.

6. Many of the challenges that remain are documented in this chapter. Future generations will have to combat not only gender-role stereotypes, but ideologies and structures that privilege men and handicap women as well. In recent years, employment-equity and equal-pay-for-work-of-equal-value policies have been developed to remedy some of the inequalities in the labour force. Analysts have also documented the various ways in which women's participation and influence

in the political arena can be enhanced. As you finish your schooling, live with a spouse or partner, hold a paying job, and become politically involved, you will experience or witness some of the gender inequalities discussed in this chapter. However, you may also witness, and work toward remedying, the gender inequalities that currently exist.

Web Links

Women in Canada 2005

http://www.statcan.ca/english/ads/89-503-XPE/index.htm

http://www.statcan.ca/english/freepub/89-503-XIE/0010589-503-XIE.pdf

OK—this site is a bit tricky to navigate, but it's worth the time. The home page will bring you to sort of an advertising site that will ask you to send money for this very excellent report, but if you click on any of the "live" 89-503-XPE links, it will bring you to a page that will allow you to view a PDF version for free. It's totally worth it as a resource that provides references for virtually all the latest info on women and what they're doing in Canada today!

Women and Politics in Canada

http://www.equalvoice.ca/research.html

This site provides current statistics and other resources to give you a greater understanding of the marginalization of women in politics. There are some great debate topics that could stem from the list of speeches and publications!

Break the Glass Ceiling

http://www.breaktheglassceiling.com

This site provides guidance and support to individuals being marginalized by the glass ceiling, as discussed in the text. What is interesting about this site is that the focus is not just on women, who are typically linked to the glass ceiling, but other minorities in society who are affected by this phenomenon as well.

Canadian Research Institute for the Advancement of Women

http://www.criaw-icref.ca/indexFrame_e.htm

This is one very happening website—and a definite must see if you're a female student with high hopes of being proactive and active in achieving women's equality through academic scholarship. Truly an amazing find!

Dads Today

http://www.dadstoday.org

Is the grass always greener? This site provides not only a resource to stay-at-home dads but also links to other support networks and ongoing conventions that a stay-at-home dad can become involved with.

Critical "Linking" Question

Take a look at the Dads Today website. I think that you'll see that it's not a very fancy website and that it looks like it could use a bit of development yet. Personally, I think that this site is a great idea. What kinds of suggestions or sites could you e-mail the Webmaster of this site to help it develop and grow?

Solutions

True or False?

1. T
2. F
3. T
4. T
5. F
6. T
7. T
8. T
9. F
10. F

Multiple Choice

1. D
2. A
3. A
4. A
5. B
6. A
7. B
8. B
9. C
10. A
11. C
12. C
13. B
14. C
15. A

Chapter 8
Race and Ethnic Relations

Chapter Introduction

Oh, what troubled times we live in! Mel Gibson made a racial slur about Jews; Michael Richards (Kramer) completely lost it and used the dreaded "N" word at a comedy festival and Isaiah Washington called his cast-mate a "faggot." Can you imagine how busy their public relations staff must be or the size of their holiday bonuses this year? Did you notice that I didn't refer to them as Christmas bonuses? That's because I'm trying to be politically correct—something that these three rich and famous celebrities could use a few lessons in. What would our world be like if everyone had the same skin colour, taste in food, and chances for employability? I think that it would be a very boring world indeed. Can you imagine how dull life would be if it weren't for the diversity that fuels our minds with wonder and feeds our soul with a plethora of music and colour? A trip to Toronto is always like a trip to another world for me. I could spend hours (and sometimes have) just walking up and down the busy streets, taking in the atmosphere and aroma of all the different cultures. I feel very proud to be living in a country that celebrates the differences among people. Although racism, discrimination, and prejudice still exist, I think that they diminish substantially with every generation and I hope that you will raise your children to appreciate the diversity that we have come to call home.

This chapter examines what sociologists mean by terms such as ethnicity, race, and racism and then discusses various theoretical approaches to the study of ethnic and racial relations. Additionally, the author investigates ethnic relations in Canada in order to show how power and resource imbalances play an important part in structuring the relationships among groups.

Jump Start Your Brain!

In this chapter you will learn that:

1. The study of race and ethnic relations involves the analysis of the unequal distribution of power and resources, and it involves a number of sociological approaches; among the most important are frustration-aggression, sociobiology, socialization, and power-conflict approaches.
2. *Ethnicity* and *race* are terms used to categorize groups on the basis of cultural and physical criteria.
3. Although the concept of race has no basis in biology, and although ethnic identities and boundaries are situational, variable, and flexible, race and ethnicity are important parts of social reality.
4. Aboriginal people in Canada are Indians, Métis, and Inuit. There are two main sociological interpretations of Aboriginal people's socioeconomic status in Canada: the culture of poverty thesis and the internal colonial model.
5. The nationalist movement in Quebec has deep historical roots. The contemporary nationalist movement is united around the goal of maintaining the French character of

Quebec. One of the main problems facing the nationalist movement in Quebec is exactly how to define the boundaries of the nation.

6. Immigration played a central role in both the early and the later phases of Canadian capitalist development. Canada accepts refugees, family class, and independent immigrants, each of whom is subject to different selection criteria.
7. John Porter's description of Canadian society as a vertical mosaic is no longer accurate.

Quiz Questions

True or False?

1. The sociology of ethnic and racial relations primarily concerns the study of how power and resources are unequally distributed among ethnic and racial groups. True or False
2. Dudley George was shot and seriously wounded by an Ontario Provincial Police officer in a chaotic altercation the day after the occupation of Ipperwash started. True or False
3. Until 1985, Indian women who married non-Indian men, along with their children, lost their federally recognized Indian status—they became marginalized Indians. True or False
4. Most of the ethnic categories that we take for granted are actually recent historical creations. True or False
5. Barker (1981) argued that the new ethnicity involves the beliefs that although some groups of people cannot be ranked biologically, with some being inferior and some superior they are naturally different from each other, and that social problems are created when different groups try to live together. True or False
6. Sociobiologists offer a popular form of primordial theory. They suggest that prejudice and discrimination—practices that deny members of particular groups equal access to societal rewards—stem from our supposedly biologically grounded tendency to be nepotistic. True or False
7. A small but nevertheless significant proportion of Aboriginal men and women work in skilled professional and technical occupations, and others are owners or managers of small and large businesses. True or False
8. In 1995, 25 percent of Quebeckers voted for separation. True or False
9. Aboriginal people in Canada are made up of Indians, Métis, and Inuit. True or False
10. Though written in 1965, John Porter's description of Canada as a "vertical mosaic" is still very indicative of the way in which ethnic groups tend to occupy different and unequal positions in the stratification system. True or False

Multiple Choice

1. A survey conducted in 1990 by Decima Research Ltd. showed that ____________ percent of Canadians agreed with the statement "All races are created equal."

   a. 60

   b. 70

   c. 80

d. 90

2. The _____________ thesis suggests that ethnic and racial attachments reflect an innate tendency for people to seek out, and associate with, others who are similar in terms of language, culture, beliefs, ancestry, and appearance.

   a. frustration-aggression

   b. essentialist

   c. sociobiologist

   d. primordialist

3. Statistical evidence shows that Aboriginal peoples are the most socially and economically disadvantaged groups in the country. In 2002–03, more than _____________ percent of housing units on reserves in Canada needed to be replaced or were in need of major or minor repairs.

   a. twenty

   b. forty

   c. eighty

   d. sixty

4. In addition to attending to the religious needs of its French-speaking parishioners, the Catholic Church acted as an agent of _____________ over French-Canadian workers and farmers.

   a. stratification

   b. change

   c. upward mobility

   d. social control

5. _______________ theories of ethnic and racial prejudices concentrate on the way in which prejudices are transmitted through socialization and the social circumstances that compel discriminatory behaviour.

   a. Normative

   b. Primordialist

   c. Power conflict

   d. Social psychology

6. No single variable can explain the complex pattern of immigration to Canada. Over the past 100 years, six main variables have influenced which groups of people have been let into the country as immigrants. Which of the following is *not* one of the variables?

   a. social class

   b. ethnic and racial stereotypes

   c. geopolitical considerations

   d. political affiliation

7. In order to enter Canada, skilled workers are selected by the federal government on the basis of their ability to meet certain minimum work experience requirements, to prove that they have enough funds to support themselves and their family members while in Canada, and merit is measured by the points system. An applicant has to earn a minimum of ____________ out of 100 points to "pass" and potentially gain admission to Canada as a skilled worker.

   a. 76

   b. 80

   c. 67

   d. 85

8. The Quiet Revolution describes the social, political, and cultural changes that occurred in Quebec in the 1960s, in part because of the initiatives of a new middle class. These changes included seven different aspects, which one is *not* one of the changes:

   a. secularization of the educational system

   b. the rise of a new francophone middle class of technical workers and professionals

   c. reform of the civil service

   d. growth in the provincially controlled public sector

9. ____________ are made up of people who identify themselves, or who are identified by others, as belonging to the same ancestral or cultural group.

   a. Marginal groups

   b. Races

   c. Ethnic groups

   d. Subcultures

10. The concept of institutional racism refers to "discriminatory racial practices built into such prominent structures as the political, economic, and education systems." Which of the following is *not* a form of institutional racism?

    a. circumstances in which institutional practices are based on explicitly racist ideas

    b. circumstances in which societal institutions such as religion and education fail to promote racial integration

    c. circumstances in which institutional practices arose from but are no longer sustained by racist ideas

    d. circumstances in which institutions unintentionally restrict the life-chances of certain groups through a variety of seemingly neutral rules, regulations, and procedures

11. Split labour market theory was developed by Edna Bonacich because of the limitations of orthodox Marxism in analyzing racism. Which of the following is *not* one of her criticisms?

    a. Marxism tends to assume that the capitalist class is all-powerful, and that other classes play no role in the development of racist thinking.

b. Because Marx's theories are largely based on economics, they can't be used as a viable tool in explaining racism.

c. Marxism portrays racism in overly conspiratorial terms.

d. Marxism has trouble explaining why racialized conflict so often results in exclusionary practices.

12. Table 8.5 on page 218 provides information on some of the important recent patterns of earnings inequality for visible minorities. The findings should lead us to be cautious about concluding that ____________ constitutes a fundamental socioeconomic dividing line in Canadian society.

    a. gender

    b. race

    c. ethnicity

    d. social class

13. ____________ argue that racism is an ideology—a set of statements shaped by economic interests about the way the social world "really works."

    a. Structural functionalists

    b. Politicians

    c. Orthodox Marxists

    d. Normative theorists

14. John Porter argued that Canada was a "vertical mosaic," a social structure in which ethnic groups occupy different and unequal positions within the stratification system. Evidence suggests that the vertical mosaic is declining in importance, at least for European immigrants and people born to Canada Discrimination against immigrants from visible-minority groups is ____________.

    a. decreasing

    b. still a problem

    c. increasing

    d. completely eradicated

15. There are three main categories of immigrants in Canada. Which of the following is *not* one of them?

    a. refugees

    b. family class

    c. adoptive

    d. economic/independent

Critical Thinking

1. The issue of multiculturalism is one that often sparks heated discussions. In many cases, marriage within your race is the only acceptable form of marriage. Some Canadians believe that race relations in Canada are getting worse, and on the other end of the spectrum, many describe the joys of living in a multi-ethnic country. Often I find that students are hesitant to admit their true feelings about Canada's immigration policies because they fear being labelled as racist or prejudiced. This learning guide is a very safe place to investigate how you really feel about this issue. Be honest with yourself and describe how you feel about multiculturalism in Canada. Have your feelings changed since you started postsecondary education?
2. In this time of political correctness, it often seems as though I'm walking on eggshells when I attend a multicultural function, but it looks as though I have nothing to worry about compared to some what some of Hollywood's celebrities are up to! What's your take on these "famous faux pas"? Should the "stars" just "fess up" and admit they made a mistake or should they continue to rely on they PR specialists to extricate them from these rather sticky situations?
3. Since 1962, ethnic and racial stereotyping in selecting new immigrants has become less important. Canadian immigration policy is now more open in terms of ethnic and racial origins of immigrants. Before 1961, Europeans made up over 90 percent of total immigrants to Canada. In 2001, immigrants from Europe made up 17.26 percent of the total flow of immigrants to Canada. Examine Table 8.3 on page 213 to take a closer look at the incredible changes that have occurred over time. How can you account for these changes? Do you think that Canada's stand on multiculturalism has affected both the number of immigrants entering Canada and also the countries from which they come?
4. The term "Quiet Revolution" describes the social, political, and cultural changes that occurred in Quebec in the 1960s, in part because of the initiatives of a new middle class. It's hard to believe, four decades later, that the politics of this situation are still unresolved and even harder to believe that Stéphane Dion has reignited this issue. In your opinion, can the issue of Quebec's separation be resolved? What are the sociological implications of separation? If Quebec becomes an independent nation, what effect will it have on you—if any?

Time to Gear Down!

At the conclusion of this chapter, you should be able to discuss or write about the following, without having to rely on the textbook:

1. Ethnic categories and identities are not fixed and unchanging; they evolve socially and historically. Canadians may now be considered an ethnic group, as they display a strong desire to define their ethnicity as "Canadian."
2. Racism refers to certain kinds of ideas and to certain kinds of institutional practices. Institutional racism refers to circumstances where social institutions operate, or once operated, on the basis of racist ideas. There are three forms of institutional racism.
3. Racism, prejudice, and discrimination have been analyzed from different sociological perspectives. Social-psychological theories, primordialism, normative theories, and power-conflict theories each offer different interpretations of ethnic and racial hostility.

4. The term "Aboriginal people" includes people who are defined in the Constitution as "Indian," "Métis," and "Inuit." The terms used to describe Aboriginal people are socially negotiated and change because of shifts in power relations among groups.
5. The culture of poverty thesis was used in the 1970s as a way of explaining the poor socioeconomic conditions of Aboriginal people. Problems with the culture of poverty thesis led to the development of the internal colonial model, a variant of conflict theory. Conflict and feminist sociologists are beginning to be more interested in class and gender diversity within the Aboriginal population.
6. French-English relations in Canada are about power relations. Material inequalities between French and English in Quebec provided the historical basis for the emergence of nationalism in Quebec. The contemporary nationalist movement has a diverse class base.
7. There are debates in the nationalist movement about who is Québécois. Tensions exist between ethnic and civic nationalists.
8. Immigration has played different roles in Canadian history. During the nineteenth century, immigrants contributed to capitalist state formation. Now, immigrants contribute to the social and economic reproduction of Canadian society.
9. There are six main variables that have shaped immigrant selection in Canada: social class, ethnic and racial stereotypes, geopolitical considerations, humanitarianism, public opinion, and security considerations. Immigrants are categorized as refugees, family class, or independents. Independent immigrants are selected on the basis of the points system.
10. John Porter argued that Canada was a "vertical mosaic," a social structure in which ethnic groups occupy different, and unequal, positions within the stratification system. Evidence suggests that the vertical mosaic is declining in importance, at least for European immigrants and people born in Canada. Discrimination against immigrants from visible-minority groups is still a problem.

Web Links

Culture.ca: Canada's Cultural Gateway

http://www.culture.ca

Explore this site to familiarize yourself with Canadian culture through various links and content, especially relating to Aboriginal People. This next link provides a direct connection to a presentation of Aboriginal Culture in Canada:

http://www.culture.ca/showcase/200604/shp001000042006e.html?wt.srch=1&gclid=CI7vz8mT-IkCFSdjQQodMRQIPA

UNI—Uniting Canada

http://www.uni.ca

This site provides a comprehensive look at the issues surrounding Quebec separatism and provides additional links to both federalist and secessionist sites. How objective do you think this information is?

Tiny Giant

http://www.tgmag.ca/index_e.htm

A very cool site! It's an organization for youths who work collectively and discuss information to disseminate concerning the seriousness of racism. A bit tricky to navigate, but it's worth the time you put into it.

The Ipperwash Inquiry

http://www.ipperwashinquiry.ca/faq/index.html

This site will bring you to the FAQ sheet of the Ipperwash Inquiry home page. All of the formal information collected during the trail can also be found on this site—read it for yourself and see who you believe!

Action Week Against Racism

http://www.inforacisme.com/en/home.php

This is an extensive site about the fight against racism in Canada that provides a wealth of information about Action Week Against Racism as well as programs, resources, and links to other important sources of information.

Critical "Linking" Question

Chances are that most of you were not even aware of the severity of the issue during the referendum of 1995 that very nearly saw Quebec separate from Canada. Visit the UNI website and see what the implications are if Quebec decides to become an independent nation within the next few years. Would it affect you?

Solutions

True or False?

1. T
2. F
3. F
4. T
5. F
6. T
7. T
8. F
9. T
10. F

Multiple Choice

1. D
2. D
3. D
4. A
5. A
6. D
7. C
8. B
9. C
10. B
11. B
12. B
13. C
14. B
15. C

Chapter 9
Inequality Among Nations: Perspectives on Development

Chapter Introduction

How often do you think of what's going on around the world? Are the commercials asking you to sponsor a child in a Third World country the only reminders that extreme poverty is prevalent in a large part of the world? How often do you find yourself in a riveting conversation about transnational corporations or supranational institutions? Well, I'll tell you, it took me a very long time to learn to look past what's just outside my windows and to use my sociological imagination to really see how the rest of the world lives—and dies. I've never been more globally aware than I am at this stage in my life and I think it's because I have more access to news through the media outlets—like my computer—but also because of the devastating world events that have been happening. I'm mindful that not all events are negative—Live 8 was amazing—Pink Floyd actually together again—but it's chiefly the horrible things that we hear about and now, or watch on YouTube and other available websites. I know that it's hard to think about children being abused when you're looking at that pair of new shoes that you simply must have, but I really hope that you do take the time and think about those children—even just for a second, before you fork out the dough to buy the shoes. It's almost certain that those children will see none of that money and yes, those shoes will look great, but at what cost? I will tell you that I am, without a doubt, the world's biggest shoe-aholic, and even I have curtailed my shopping in this area, specifically because of how the products are made—and it's not just shoes. You're a bright person, or you wouldn't be reading this guide. The next time you go shopping, take the time to find out where your potential purchases come from—you can do this before you head out—and I guarantee that you'll feel a squillion times better when you buy items that you know have been made by people who have not been exploited, or that have not been tested on animals. Trust me—would I steer you wrong?

This chapter allows you to look at those countries from a developmental perspective and gives you an understanding of why the poverty remains. By exploring the competing perspectives on why there are rich and poor countries, we gain a greater understanding of what can be done to enable everyone to fully develop their own unique capacities.

Jump Start Your Brain!

In this chapter you will learn that:

1. Transnational corporations are prepared to shift worksites to maximize profits. This undermines the employment and well-being of many people.

2. Two hundred and fifty years ago there were small gaps in living standards and levels of productivity between countries. Now the gaps are huge.
3. The poor countries today are mainly those that were colonized by the West.
4. Liberal development theory argues that if markets are allowed to function without state interference, prosperity follows. Although liberal theory has been adopted by some countries, it has not led to improvements in the well-being of most citizens of less developed countries.
5. Dependency theory suggests that underdevelopment results from foreign economic and political control. However, the development breakthroughs of some countries suggest that domestic control of corporations and activist governments can be the keys to economic success.

Quiz Questions

True or False?

1. Anti-imperialist perspectives are the mirror-image of neoliberal views. Proponents argue that most countries are poor because Western capitalism and imperialism have penetrated them too deeply. True or False
2. Transnational corporations are prepared to shift the site of work in order to maximize profits. This has created massive favourable opportunities for peoples' employment and well-being. True or False
3. Handicrafts still accounted for half of world industrial output in 1900. Capitalist industrialization and empires brought the less developed countries to levels of prosperity not experienced for centuries. True or False
4. Liberalism is the classical doctrine that equates capitalism with freedom and distrusts the power of governments. Liberalism is a doctrine that removed the power of kings and aristocrats in the 1700s and now justifies the power of the few families and foreign corporations that dominate Canada's economy. True or False
5. Economists often measure growth and living standards by using the gross national product (GNP) or gross domestic product (GDP) data. For the most part, these terms are interchangeable. GNP is the value of all goods and services produced in a country in one year as measured officially by money transactions. True or False
6. Colonialism involves one country taking over another by force and ruling it without consent. In informal colonialism, a country gains control over another country through diplomacy, economic threats, and incentives, and, as a last resort, threatening or carrying out regime change militarily. True or False
7. Disparities between rich and poor countries are enormous. Life expectancy is 65 years in the less developed southern hemisphere compared with 79 years in rich capitalist countries. True or False
8. The weakness of modernization theory is in its ethnocentric, evolutionary assumption that all countries must develop in different stages than Western countries. True or False

9. A child born in the United States or Canada will consume, on average, 10 times the resources and produce 10 times the pollution of a child born in Bangladesh or Bolivia. True or False

10. For Marx, capitalism contains "the seeds of its own destruction." As capitalism replaces less efficient modes of production, such as feudalism, and destroys entire classes, such as serfs, it creates larger and larger communities of wage earners who have to sell their labour power to survive. True or False

Multiple Choice

1. In 1750, most children in London did not live to see their ____________ birthday.

   a. fourth

   b. fifth

   c. sixth

   d. seventh

2. The Industrial Revolution had its roots in the ____________ industry. It was a peculiar industry to transform the world.

   a. forest

   b. mining

   c. agricultural

   d. cotton

3. The ___________ was set up in 1944 to facilitate international trade and corporate investment by making national currencies readily convertible into the currencies of other member countries.

   a. IMF

   b. World Bank

   c. EU

   d. OECD

4. ________________ profoundly affects a person's life chances everywhere and especially in less developed countries.

   a. Age

   b. Socioeconomic status

   c. Gender

   d. Race

5. In the eighteenth and nineteenth centuries, ____________ was the most remote of the three fabled "rich" countries of Asia that were coveted by the West.

   a. Japan

b. Hong Kong

c. Singapore

d. Taiwan

6. In _____________, a major oil exporter, the lending boom ended when world oil prices fell in 1982 and they could no longer pay debt charges.

a. India

b. Mexico

c. Saudi Arabia

d. Zambia

7. Pharmaceutical companies spend billions of dollars on research if they anticipate big profits protected by patents. But corporations do not profit from environmental protection and wiping out poverty, illiteracy, and infectious disease. Consequently, more than _______________ children a year die needlessly because they lack simple low-cost interventions, such as immunization against infectious disease.

a. 5,000,000

b. 1,000,000

c. 2,500,000

d. 6,000,000

8. Experts disagree about the answers to the question of perpetuating global inequalities. Sociologists may first distinguish two broad approaches to answering them: neoliberal and “anti-imperialist.” Neoliberalism is the dominant approach in _______________ and other Western countries.

a. Canada

b. United States

c. Germany

d. France

9. All is not always what it appears to be. If average income rise, do most people benefit? Not necessarily. National averages hide poverty by lumping together incomes of rich and poor people. For example, Canadian billionaire Kenneth Thomson was the world’s ninth-richest person, with assets of _______________ before his death in June 2006.

a. US$4.8 billion

b. US$15.5 billion

c. US$17.9 billion

d. US$22.2 billion

10. Domestic control of corporations and activist governments were keys to the historical success and spectacular development breakthrough of _____________ and they are likely to benefit less developed countries today.

    a. Japan and Sweden

    b. South Korea and Sweden

    c. Singapore and Japan

    d. Sweden and Hong Kong

11. Latin America used to be home to dependency theory, inwardly directed development, and which other prevailing ideological concept?

    a. neoliberal/Westernization perspective

    b. modernization approach

    c. dependency theory

    d. Marxism and dependency theory

12. Since _____________, Singapore, Hong Kong, South Korea, and Taiwan, the four Asian "Tigers" have gone from underdeveloped to developed status.

    a. 1945

    b. 1950

    c. 1960

    d. 1930

13. With 10 percent of the world's people, Africa suffers more than _____________ of the world's conflicts.

    a. 40%

    b. 20%

    c. 80%

    d. 50%

14. Poor pregnant women face enormous health risks. Poor women's risk of dying during pregnancy and birth is up to _____________ times higher than in developed countries.

    a. 200

    b. 400

    c. 600

    d. 800

15. The globalist "Washington consensus" mandates every country to remove domestic control over its own economy and adopt neoliberal economic policies. In exchange for loans and financial aid, such policies require that countries follow seven guidelines. Which of the following is *not* one of them?

a. dismantle laws against foreign ownership and control

b. cut public expenditures on health care, education, and the like

c. tax farm families to increase GNP

d. set currencies at low levels to encourage exports

Critical Thinking

1. It's becoming virtually impossible for me to shop, because I boycott so many stores and products. Admittedly, some of my boycotts are ridiculous—I don't eat Captain Crunch cereal because his eyebrows arc on his hat, not his head—but I also refuse to use products that have been tested on animals. Over the past few years, I've made a concerted effort to be more aware of where the products and services I use originated, and I'm very careful not to support products that have been made in "sweat shops" or in countries where I know that the labour is working in poverty-stricken conditions. Doing this requires me to be, at the very least, a little globally aware of transnational corporations and the far-reaching implications of them. How would you rate your own global awareness? Do you read the labels before you purchase items? Would it bother you to know that a sweater you were wearing was made by children who were being forced to work for pennies a day? Do you boycott certain stores or items? If so, which ones?

2. Decisions, decisions. Pretend that you have to participate in a debate entitled "Why are some countries rich and others poor?" You get to choose which side you're on but one side addresses the issue from the neoliberal perspective and the other side views the situation from the anti-imperialist paradigm. Neoliberalism argues that poor countries can better themselves by becoming more like "us" and they remain poor because they're lazy and have corrupt elites. Anti-imperialist doctrine espouses the belief that rich countries remain rich because they exploit the poor countries. Which side of the debating team would you choose to be on and why? What examples would you use to defend your team and ultimately win the debate club MVP award?

3. World markets do not work according to textbook formulas. Real markets are dominated by a few giant companies that use monopoly positions to capture excess profits. Controlled by dominant capitalist countries, the IMF and the World Bank determine the policies of Third World governments. Instead of allowing governments to develop their own repayment policies, the IMF imposes rigid prescriptions that benefit Northern bankers and local, often corrupt, elites. What appears be extortion is tolerated in our society today. Who, if anyone can put an end to this? As an individual, do you feel as though there is anything you can do to help?

4. In 1999 the World Bank fired Joseph Stiglitz, the Bank's chief economist because he expressed doubts about World Bank and IMF neoliberal policies. The World Bank and the IMF, though separate institutions, are linked by "triggers." For example, taking a World Bank loan to build a school "triggers" a requirement to accept every conditionality—on average, 111 per nation—laid down by both the World Bank and IMF. Loans and assistance packages to countries in economic

trouble are, says the World Bank, designed after careful in-country investigation. Joseph Stiglitz disagrees and told an investigative journalist that the "investigation" consists of close inspection of a country's five-star hotels. Though he did lose his job, in 2001 he won the Nobel Peace Prize for Economics. If you were Joseph Stiglitz, would you have made the same decision he did? What are the implications of his actions?

Time to Gear Down!

At the conclusion of this chapter, you should be able to discuss or write about the following, without having to rely on the textbook:

1. The search for profits leads transnational corporations to move worksites to places where labour costs are low. This has negative implications for the well-being of many people.
2. Great disparities exist in well-being of rich and poor countries, as well as between classes and genders within all countries.
3. In 1750, substantial gaps in living standards and levels of productivity among countries did not exist. For the most part, today's poor countries are those that were once colonized by the West.
4. The assumptions of liberalism about free markets and capitalist development have been adopted by the International Monetary Fund, transnational corporations, and the U.S. government. They do not live up to their promise to improve the well-being of most people in less developed countries.
5. Anti-imperialist theories emphasize the importance of domestic economic control and democracy in the development of the less developed countries.
6. Domestic control of corporations and activist governments were keys to the historical success and spectacular development breakthrough of Sweden, and they are likely to benefit less developed countries today.
7. Domestic control of corporations and activist governments were the keys to success in the spectacular development breakthroughs that mark the histories of South Korea and Sweden.
8. More democratic, egalitarian, and supportive policies for development will benefit all countries. Kerala is a good example of this. Those in the rich North may have to sacrifice some material wealth to share more around the world and ensure ecological sustainability.

Web Links

IDRC-CRDI

http://www.idrc.ca/en/ev-1-201-1-DO_TOPIC.html

Awesome website! The International Development Research Centre is an excellent agency that assists developing nations with their own strategies to learn about and overcome their own levels of poverty. Take the time to read some of their publications and funding opportunities—there's a wealth of information here—no pun intended!

The United Nations (UN)

http://www.un.org

A very happening website and totally up on the graphics! I checked out the children's page and it's got cartoons! Check out the "UN Millennium Development Goals." Do you think that they're realistic and attainable? To what extent?

Global Policy Forum

http://www.globalpolicy.org

This forum monitors policy making at the United Nations, and has daily updates to keep you apprised of current statistics and real issues relating to global inequality in developing nations. Check out the "Social and Economic Policy" link and you will find information on poverty and development, world hunger and global injustice and inequality.

IMF—International Monetary Fund

http://www.imf.org

The IMF is an organization that works together to promote global monetary cooperation and reduce poverty.

The World Bank Group

http://www.worldbank.org

This website offers an abundance of information about global equality and inequality from the viewpoint of one of the world's largest moneylenders to developing nations. How objective do you think this information is?

Critical "Linking" Question

This question involves a little bit of work. Take a good look at the IMF website and compare it to the World Bank website. There are certainly a lot of commonalities, but what do you think the major differences between these two groups are?

Solutions

True or False?

1. T
2. F
3. F
4. T
5. T
6. T
7. F
8. F
9. T
10. T

Multiple Choice

1. B
2. D
3. A
4. C
5. A
6. B
7. D
8. A
9. C
10. A
11. C
12. C
13. A
14. C
15. C

Chapter 10
Families

Chapter Introduction

Whenever I begin teaching, I introduce myself to my students and almost always tell them how lucky I am to have such a wonderful and supportive family. This doesn't usually get a reaction until I tell them that I'm the second-youngest of thirteen children! Big families are not very common these days, and I guess you can say that like many familial arrangements in society, my family doesn't fit into the neat little definition of the nuclear family that has been around for decades. I can't imagine a life without brothers or sisters. I have six sisters and originally I had seven brothers, but two have joined my Dad in heaven, so there are eleven of us left. Let me tell you that Christmas is a wild scene! Even though I have a very large family and some of them live sort of far away, the miles between us have not added distance—we are all very close. And though we certainly wouldn't be defined as a traditional family, each of us is very traditional in the sense that we value the very basic beliefs in honesty, respect, and trust. Though the gender division is alive and well today, it wasn't very prominent when I was growing up, mainly because everyone had their chores to do, and gender just didn't fit into the picture. All of my brothers can still iron a shirt (probably better than I can) and they all know how to cook and clean—again better than me—except maybe the cooking, because that's one area that I'm especially good in. We all worked part-time jobs—I started picking fruit when I was eight years old—and the money all went into the family pot, in order to help my parents feed and clothe us. It didn't dawn on any of us to ask for an allowance because we just knew that the money wasn't there, and yet as I look back, I don't feel as though I was cheated out of anything as a child. At Christmas time, the presents would barely fit under the tree and they were quality gifts—and not only ones that were homemade. But, as you'll see in this chapter, the ways in which we have come to define the "traditional" family has changed over time, and I think that that's a good thing.

In this chapter, the author reviews the dilemmas that many families face in the light of the popular myths that surround them. Because family life is so familiar to us, we often accept the theories that explain it and the commonsense understandings of it without a second thought. Most families experience a variety of problems that are often swept under the rug and explained away by the personal, private characteristics of "human nature." In this chapter, you'll read about two patterns of family life that are very different from our own. In foraging societies, families lived in a communal setting and were based on reciprocity, where everyone shared the family chores. The social relations of the agricultural families were also the relations of production, where the chief economic relationship was that between the husband and wife. By the end of the chapter, you'll have learned how families have changed over time and evolved into the different family forms that we have in today's society.

Jump Start Your Brain!

In this chapter you will learn that:

1. Although common sense suggests that current dilemmas in family life are private problems, sociology helps uncover their public sources and solutions.
2. The family does not have a universal form and its structure is therefore not a product of some biological imperative: families vary widely in the way they are organized across different cultures and through history, and family organization is loosely related to the way material production is organized.
3. Our society is organized around a gendered division of labour and the heterosexual nuclear family; this is the main reason that families that assume a different form—especially lone-parent families—have a particularly hard time meeting the needs of their members.
4. Although women have increasingly assumed part of the financial support of their families, men have not come to share the work that must be done in the home, nor has society changed in ways that accommodate this changed reality for families.
5. The gendered division of labour that makes it possible for nuclear families to care for young children involves sizable liabilities for women and children. The social isolation of full-time mothers and the stress attached to their high-demand, low-control situation reduces the quality of child-care.
6. The chief negative effect of divorce for women and children is the loss of income that follows. The most effective solution to this problem—government support of all children—has not been adopted by the Canadian government, although most advanced industrial societies do have such a policy objective.

Quiz Questions

True or False?

1. Although conventional patterns are in decline, our society still seems to be organized around traditional nuclear families and the assumption that children are a private responsibility. True or False
2. Lesbians and gays face fewer challenges creating families in today's society given the heterosexual nuclear family is no longer the popular fixation. True or False
3. The chief negative effect of divorce for women and children is not the loss of income that follows—but rather—the emotional upheaval that is difficult to transcend. True or False
4. When "mothers' allowances" were introduced in early-twentieth-century Canada, they were paid only to widows even though other types of lone mothers were equally in need of assistance. True or False

5. A key problem with the social conflict perspective is its focus on how institutions create social order, and its consequent failure to analyze the tensions in family life that can generate social change. True or False
6. Sexual behaviour in same-sex couples also reflects gender differences. Typically, gay male couples have more frequent sex, and lesbian couples have less sex, than heterosexual couples. True or False
7. Agricultural societies are societies in which people acquire their subsistence from the resources around them, without cultivating the earth. True or False
8. The gendered division of labour that makes it possible for nuclear families to care for young children involves sizable liabilities for women and children. The social isolation of full-time mothers and the stress attached to their high-demand, low-control situation reduces the quality of child-care. True or False
9. It is estimated that about 30 percent of marriages in Canada will end in divorce, which is considerably high compared to rates in the United States and Sweden. True or False
10. Our family patterns developed out of the patterns that were typical of precapitalist agricultural societies. In those societies, the household itself was the productive unit; producing subsistence was its main objective. True or False

Multiple Choice

1. Only in the _____________ century did people begin to assume that romantic love, sex for the sake of pleasure, and marriage should be intimately bound together.

   a. 17th

   b. 18th

   c. 19th

   d. 20th

2. In _____________ people acquire subsistence by gathering edibles and hunting live game.

   a. preindustrial agricultural societies

   b. matrilineal societies

   c. foraging societies

   d. egalitarian societies

3. A _____________ has sustained families in an economy in which employers bear no direct responsibility for the welfare of their employees' families.

   a. structural functionalist perspective

   b. family wage

   c. nuclear family

d. gendered division of labour

4. _____________ hold(s) that, as with physical traits, social behaviour is inherited biologically—in other words, that behaviour can be linked to specific genetic configurations.

   a. Eugenics
   b. Pronatalists
   c. Sociobiologists
   d. Biological determinists

5. In terms of family diversity, cohabiting men and women constitute the fastest-growing type of family in Canada. They more than doubled in number between 1981 and 1991; and by 2001 almost _____________ of families involved a common-law couple.

   a. 15%
   b. 14%
   c. 50%
   d. 18%

6. Although most European countries provide more generous state support of families than Canada, _____________ has especially good policies.

   a. Switzerland
   b. Finland
   c. Sweden
   d. Denmark

7. _____________ is the argument that individual behaviour or social organization is directly caused by biology or biological processes.

   a. Sociobiology
   b. Evolutionary psychology
   c. Pronatalism
   d. Biological determinism

8. Since the late 1990s, _____________ has moved in the opposite direction—toward creating a broad set of family policies that feature universal, affordable child-care.

   a. Ontario
   b. British Columbia
   c. Quebec
   d. Alberta

9. It is widely believed that the best unit in which to raise children in the nuclear family. This commonsense notion was promoted by ____________.
   a. Emile Durkheim
   b. Talcott Parsons
   c. Karl Marx
   d. George Herbert Mead
10. Although structural functionalism dominated the study of families until recently, there are obvious problems with this perspective. Which of the following is *not* one of them?
    a. Just because an institution performs a social function, there is no reason to assume that some other institution might not perform that function equally well.
    b. Its focus on how institutions create social order and its consequent failure to analyze the tensions in family life can generate social change.
    c. The functions that are emphasized allegedly meet the needs of society, but not necessarily, the individuals in it.
    d. It is predicated on the historical definition of the nuclear family, which is, for the most part, outdated and not indicative of current trends in Canadian society.
11. Whereas the absence of privatized family life empowered individuals in foraging societies, private ownership of the means of production and a considerable struggle to survive meant that agricultural societies had certain characteristics. Which is not one of them?
    a. All individuals were subordinated to the household enterprise.
    b. Elders held power and made most decisions for the clan.
    c. Women were subordinated to men.
    d. Children were subordinated to parents.
12. In 2001, ______________ of families consisted of one parent—usually a mother—and her dependent children.
    a. 15.7%
    b. 17.5%
    c. 18.9%
    d. 22.4%
13. The trade-union movement responded to the straits of working-class life with a campaign for a ______________; it refers to a wage that is paid to a man and is sufficient to support him, his wife, and his children.
    a. decent wage

b. family wage

c. minimum wage

d. wage based on commission

14. Marriage involves negotiation—if not daily, then at least periodically and especially early in the relationship. Women's bargaining power in those negotiations is undermined by four factors; which of the following is *not* one of them.

a. women's disadvantage in the labour market

b. women's perceived disadvantage on the remarriage market

c. a cultural devaluation of caring work

d. full-time mothers have no "downtime" and often minimal sleep

15. The ____________ of domestic labour, in effect, allows the community to avoid solving the child-care problem.

a. gendered division

b. evolution

c. privatization

d. relatively low cost

Critical Thinking

1. "Although individuals make choices about whether to marry, have children, and so on, the social forces guiding them toward those decisions are so powerful that most individuals end up making the same choices." Do you agree or disagree with this statement? What if you're a woman who chooses voluntary childlessness, or a closeted gay in a heterosexual relationship? Do the pressures to conform that are pervasive throughout society wreak havoc on your own personal beliefs and desires? How often do you "crumble under pressure" and do something because doing what you really want to do would just take too much time and effort—not to mention the emotional and possibly the financial energy. What does this statement say about individuality?

2. Research on various aspects of personal development shows that there are no significant differences between children of gay and lesbian parents and children of heterosexual parents. Do you find this result surprising? Why or why not? Do you think that children who have same-sex parents will face more adversity in life than children with heterosexual parents? Do you think that there would be different challenges if a child's parents are female or male? If so, what do you think these challenges would be and why?

3. There is a growing conviction that being a full-time mother is bad for women and, consequently, for their children, and that men's physical absence from the home can also be detrimental to the social development of children, if their absence leads to emotional distance from the children. In the past, women who worked

outside the home and left their children to daycare providers were chastised. Can valid arguments be made for both sides and, if so, what are they? With the increasing need for both parents to work outside of the home in order to live a comfortable life, is there a reasonable solution to this situation? If so, what is it? What if the government decided to pay a fair wage for household labour? Would that be a viable solution?

4. Celebrities have certainly been adding to the confusion when trying to define the already difficult to define "traditional family." Madonna, Meg Ryan, Angelina Jolie and Brad Pitt, Rosie O'Donnell, and Tom Cruise and Nicole Kidman all have adopted children from different races. Why do you think these celebrities don't adopt children from their own country? Is it just a matter of money or is there more to this story? Do you care, as long as these babies are well taken care of?

Time to Gear Down!

At the conclusion of this chapter, you should be able to discuss or write about the following, without having to rely on the textbook:

1. Commonsense arguments hold that current dilemmas in family life, such as how women can balance the responsibilities of family and paid work and how men can succeed at breadwinning and also do their share of the housework, are private, individual problems. Similarly, family-values advocates argue that change by individuals is the solution to family problems. In the absence of a sociological analysis, the public sources of these problems—and of their potential solutions—remain unclear and unrecognized.
2. Biology does not produce the family, which varies in its organization considerably across different cultures and through history. Although the nuclear family unit is common, it is not always responsible for child-care; often, the family is embedded in a larger household or community that collectively assumes responsibility for all the children of the group.
3. Family organization can be seen as loosely related to the organization of production, especially if family is defined as the sets of relationships that people create to meet the daily needs of adults and children.
4. In foraging societies, the nuclear family is embedded in a larger group that cooperates with respect to subsistence, consumption, and child-care. Paradoxically, the communal nature of these societies grants considerable autonomy to the individuals living in them.
5. In the agricultural societies of preindustrial Europe, households were primarily units of production in which the need to survive took precedence over all else. Household composition, even the texture of emotional life, reflected economic pressures.
6. Contemporary family patterns are the product of a particular history. Our history is marked by the development of an economy outside the household, in which the relations of paid employment are separated from the relations that provide for

daily personal needs and the needs of children. A gendered division of labour corresponds to this separation.

7. As women have increasingly come to share the burden of family financial support, men have not proportionally increased the work they do in the home. That jobs are usually geared to people who lack family responsibilities is partly what prevents men from taking on more housework—their time is too limited. However, ideas about gender also make men reluctant to do "women's work."
8. The gendered division of labour that makes it possible for nuclear families to care for young children involves significant liabilities for women and even for children. The social isolation that full-time mothers experience, combined with the stress attached to their high-demand, low-control situation, reduces the quality of child-care they are able to provide. Now that so many mothers are working outside the home, however, the problems associated with privatized responsibility for child-care may prove too burdensome, and government supports are likely to provide the only viable solution.
9. Because of fairly high rates of divorce, and an increasing incidence of births to unmarried women, many Canadian children will spend some part of their lives in lone-parent families. The most problematic thing about this type of family is its typically low income and the attendant stress on the parent.
10. The policies of the Canadian state pertaining to families are premised on the assumption that the welfare of family members—even children—is not the responsibility of the government or of the community. Accordingly, family law in Canada now views marriage as the union of two individuals who are responsible for their own support, even in the case of a divorcing woman who was a full-time homemaker.

Web Links

The Vanier Institute of the Family

http://www.vifamily.ca

Their vision is to "to make families as important to the life of Canadian society as they are to the lives of individual Canadians." I think that this is a very cool vision statement. Visit this website and see if you agree with how they define the Canadian family!

Parenting After Divorce

http://www.justice.gc.ca/en/ps/pad

This government site provides information mainly for parents on the judicial issues surrounding divorce, but also provides information, research, and reports relating to the possible affects of divorce on the family in Canada.

Salon

http://dir.salon.com/topics/family/index.html

Take a Hop and Stop—on this Ab-Fab site! Postmodern chic with an academic twist and lots of fun, fun, fun! Family angst and social stigma are new tickets to fame and fortune but this site has all kinds of information for you to get lost in.

Ontario Association of Children's Aid Societies

http://www.oacas.org

Remember when it was the rage to threaten your parents with calling Children's Aid? Well, this site is no joke and offers the straight dope on the welfare of children and families in Ontario. OACAS acts as a government liaison and develops policy and social programs and services to support parents and children with leading a healthy lifestyle.

Families Online Magazine

http://www.familiesonlinemagazine.com

This site is a free online magazine that acts as a resource for parents on a variety of parenting and family-related issues, such as health, recreation, education, and finances.

Critical "Linking" Question

Get your board and go surfing on the *Families Online Magazine*. I'm wondering whether you can tell if it's sexually stereotyped in any way? How do you think that the sponsors of this website have chosen to define the nuclear family?

Solutions

True or False?

1. T
2. F
3. F
4. T
5. F
6. T
7. F
8. T
9. F
10. T

Multiple Choice

1. D
2. C
3. D
4. C
5. B
6. C
7. D
8. C
9. B
10. D
11. B
12. A

13. B
14. D
15. C

Chapter 11
Work and Occupations

Chapter Introduction

I'm laughing as I type this, just thinking about all the crazy jobs that I've done in my life in order to make money to pay my tuition over the years! You couldn't pay me enough to go back to work at a factory. I did two summers working on the assembly line at General Motors, and on some days I think that I still smell like auto parts. What became very clear to me during those summers is that there is a huge gender gap in much of the work that is done in today's society—female summer students were definitely treated differently from male summer students and all summer students were treated differently from all permanent workers. One summer, I think I lasted two days as a server—back then I was called a waitress—and I hope that the poor man that I spilled hot coffee on won't be reading this. I also learned a very valuable lesson during that time—that I won't do a job that I don't like, regardless of the salary that I can earn. I am certainly not saying that this should apply to anyone other than me, but I think that it is important to determine these kinds of job characteristics before you actually start training for a career. I often ask my students how many of them would shovel manure if it paid a lot of money. I qualify it by telling them that the job would consist of eight hour shifts, with breaks and would pay an hourly wage of $78. What do you think the reaction is? Would you take this job? I am never surprised by the results. At first, literally every student indicates that they would take on this job, but when we discuss the actual work, many change their answers. I suspect that those students that say that they'd shovel manure for the rest of their lives might just end up doing that. The others will most likely invest in their education and not have to worry about falling in a pile of manure and coming up smelling like a rose. I've learned the hard way that life is too long to be doing something that you don't like to do. Confucius really did say it best: "Find a job you love and you'll never work a day in your life!"

In this chapter, you'll learn that the study of work and occupations is a study of both the constraints and the struggles that happen every day in Canadian workplaces. You'll also learn what the service economy means for our working lives, how employees and employers do battle to gain the upper hand in determining what jobs will be like, what factors lead to "good jobs" or "bad jobs," and whether these jobs will be satisfying to workers.

Jump Start Your Brain!

In this chapter you will learn that:

1. Work in Canada is dominated by the service economy. Jobs are becoming polarized into good and bad jobs, and nonstandard or part-time jobs are becoming more prevalent.

2. The labour market is segmented into areas of good and bad jobs: job ghettos in the labour market trap certain groups of workers, such as women and members of visible minorities; labour-market shelters help some workers protect the access to better jobs in the economy.
3. The degree to which technology enhances or degrades jobs is contingent on the goals of management, the type of technology is used, and the way workers react to the technology. Debate has emerged over whether computers increase productivity.
4. Management uses a variety of strategies to organize work. From scientific management to Japanese management, these strategies attempt to help management reduce costs and increase the productivity of workers.
5. Working in organizations presents specific challenges to women and members of minorities. Female managers may be perceived as less effective than their male counterparts.
6. Job satisfaction measures how workers feel about their jobs, and the characteristics of organizations and jobs are the primary determinants of satisfaction. If employers provide workers with challenging jobs, opportunities for advancement, and adequate pay, workers are more likely to be satisfied.
7. Alienation is a structural condition reflecting workers' lack of power over their work and lives. Several types of workplace behaviour, such as sabotage, game playing, and strikes, are typical reactions to alienating conditions.

Quiz Questions

True or False?

1. Most researchers speak about changes in the economy and the world of work as revolutions. Starting in England in the late eighteenth century, the economic and social structure shifted from feudalism to capitalism. True or False
2. By breaking jobs into their smallest components and removing the need for workers to make decisions about their work, Taylorism opened the door for management to reduce their reliance on skilled labour. True or False
3. The labour-market-segmentation theory suggests that regardless of where you enter the labour market, there is always an opportunity for getting a different, better-paying job. True or False
4. Job ghettos are parts of the labour market that prevent certain groups of workers from experiencing upward mobility. Structural barriers based on stereotypes work to keep some individuals from entering the primary labour market and the best jobs. True or False
5. A "great transformation" in social and labour organization led to the rise of a rural capitalist class and working class. True or False
6. The consensus that seems to be growing among researchers is that no real differences exist between male and female managerial behaviour and

effectiveness. This body of research suggests that, rather than defining certain managerial styles as male and female, we should develop a contextualized understanding of managerial styles. True or False

7. Work and organizational characteristics are the secondary predictors of job satisfaction. True or False
8. Those of us who live in the major urban industrial areas of Canada are more likely to end up in a better job simply because there are more such jobs in our regional labour markets. True or False
9. In the immediate future, we can expect employers to continue looking for ways to reduce operating costs. True or False
10. Labour-market segregation involves the separation of the labour market into sectors of high- and low-paying jobs. True or False

Multiple Choice

1. ______________ rewards are the material benefits of working. Adequate pay, benefits, and opportunities for advancement are examples.

   a. Intrinsic

   b. Extrinsic

   c. Merit-based

   d. Annual

2. In Marxist theory, ________________ is a structural condition of "objective powerlessness." Workers do not have power or control over their work situation and are separated from the means of production. This situation is indicative of work in a capitalist economy.

   a. class consciousness

   b. cultural hegemony

   c. alienation

   d. exploitation

3. _____________ is the catchphrase for programs designed to reduce barriers for women, members of visible minorities, Aboriginal people, and people with disabilities.

   a. Cultural awareness

   b. Multiplicity sensitivity

   c. Managing diversity

   d. Interculturalism

4. In Canada between 1971 and 2001, the number of women managers increased from 6 to 35 percent of all managers. However, since only ______________

percent of senior managers in Canada are women, the "glass ceiling" for women has by no means disappeared.

a. 25

b. 21

c. 30

d. 35

5. To respond to changes in demand for their products and services, some employers now rely on temporary workers, ranging from clerical help to computer programmers, hired through temporary employment agencies. In 2003, almost 13 percent of Canadian workers were employed in temporary or contract positions. Like part-time workers, temporaries tend to be ____________.

   a. young and male

   b. young and female

   c. visible minorities

   d. retired women

6. Good jobs should provide ____________ rewards such as decision-making opportunities, challenging non-repetitive work, and autonomy that allows for self-direction and responsibility over work tasks.

   a. intrinsic

   b. extrinsic

   c. merit-based

   d. annual

7. For employers, ____________ can reduce labour costs, since part-time and temporary workers receive less in pay and benefits.

   a. labour market segregation

   b. migrant workers

   c. nonstandard jobs

   d. workplace equity

8. Although we still see vestiges of the first and second industrial revolutions in our working lives, many things have changed. Most prominent is the movement away from a manufacturing-based economy toward a(n) ____________ economy.

   a. service

   b. technological

   c. knowledge

   d. agrarian

9. Between 1987 and 2003, ________________ experienced the greatest growth in employment.

   a. accommodation and food services

   b. health care and social assistance

   c. educational services

   d. service industries

10. Relating to job ghettos, today, ______________ can keep qualified minorities from being hired.

    a. labour-market segmentation

    b. employer prejudice

    c. labour-market shelters

    d. hidden job markets

11. Recent "humanization" of work and worker participation efforts witnessed quality-of-work-life (QWL) programs being introduced into workplaces. QWL is meant to accomplish three goals. Which of the following is *not* one of the goals?

    a. democratizing the workplace

    b. raise quality-control circles

    c. giving workers more responsibility and control over their work

    d. provide more complex jobs for workers

12. The "Hawthorne studies" started the movement to consider how employers could improve human relations. Which of the following is *not* an objective of the "Hawthorne studies"?

    a. employers ensure fair and equitable pay increases

    b. employers fulfill employees' social needs

    c. employers increase employees' satisfaction

    d. employers make employees' feel better about their jobs

13. According to Max Weber, bureaucracies are the most efficient and rational organization form for reaching the goals of capitalism. Which of the following is *not* characteristic of bureaucracies?

    a. Written rules provide guidelines for handling routine situations.

    b. A complex division of labour ensures that workers know what is required of them and helps to identify who is responsible when something goes wrong.

    c. Bureaucracies are a way to overcome arbitrary decisions and corruption in non-bureaucratic organizations.

    d. Bureaucracies experience increased profits because of reduced overhead costs.

14. Unions play an important role in facilitating collective action by workers. Today, _____________ of the non-agricultural paid labour force are in unions.

    a. 18%

    b. 27%

    c. 31%

    d. 41%

15. How do Canadians feel about their jobs? According to a 2001 survey, _____________ of Canadians reported that they were satisfied with their jobs.

    a. 26%

    b. 45%

    c. 74%

    d. 93%

Critical Thinking

1. The story of Debora De Angelis that begins this chapter strikes a chord of familiarity with me. Having to depend on a variety of part-timc and sessional teaching positions to finance my education, I know all too well the differences that exist between full-time employees and part-time or contractual help. Do you find yourself in the same situation as Debora? Are you overworked in an underpaid environment? Quitting a job is a big decision, especially when your job is one that allows you to earn some money while you're in school. What can you do to improve your situation? Would you support a union? Why or why not?
2. Thinking back to my summers at GM, I remember that, at the time, the economy was good and the company offered full-time positions to summer students. I remember my supervisor asking me if I'd like to stay on, and I vividly remember my response: "I'd rather eat a bug." There was not enough money in the world to make me stay on. Would you take on a job that you know you wouldn't like just because the money is good? How long do you think that you would last at it? Is there a job that you wouldn't do, regardless of the pay?
3. Until recently, only doctors had the authority to deliver babies. Doctors maintained their authority because of their monopoly over the relevant body of knowledge and used their national associations to lobby provincial and federal governments to deny others the right to practise medicine. As a result of current public demands for access to midwives and the development of professional midwifery schools, doctors have lost some control over the birthing process and have lost their monopoly over this body of professional knowledge. What do you think of this? Should midwives be covered by the same provincial medical insurance as doctors? What about those much-needed medical professionals like chiropractors or naturopaths that seem to fall between the cracks? How would you decide the membership of a professional organization?

4. Recently, I've been on this rant about customer service. It all started when I was at the checkout and for some reason the cash register didn't work, and the clerk actually had to figure out how much change she owed me in her head. Like, she actually had to do the math. Well, I have to tell you that I am not a math genius, but even I could figure it out faster than this poor girl who was searching frantically for a calculator. I felt bad for her at first, but then I thought it was awful that she had to rely on the machine to tell her how to count change. When I offered her the answer, she became quite snippy and told me she could do it on her own. Well, needless to say, there was a bit of a scene, and now that store is on my boycott list. When you're at your job, how far are you willing to go to help out a customer? Do you go that extra mile or do you just do the bare minimum of your job requirements to get by? What happens if your job requirements change? For example, you have to learn how to use a new piece of technology—a cash register for instance? Do you expect to be compensated for it?

Time to Gear Down!

At the conclusion of this chapter, you should be able to discuss or write about the following, without having to rely on the textbook:

1. In the first industrial revolution, large segments of the population moved from being peasant farmers to being wage-earning factory workers living in urban areas. During the second revolution, companies increased in size and developed administrative offices with a complex division of labour.
2. The rise of the service economy is changing the types of jobs available in the labour market. The types of jobs available in the service economy are being polarized into good jobs in upper-tier industries and bad jobs in lower-tier industries.
3. The proportion of Canadians employed in nonstandard jobs is growing. Much of this job growth is fuelled by the expansion of the lower-tier service sector.
4. Labour-market segmentation shows that different segments exist in the labour market. Good jobs are located in core industries and firms with primary labour markets, while bad jobs may be found in peripheral industries and firms with secondary labour markets. Job ghettos are areas of the labour market that trap disadvantaged groups of workers. Labour-market shelters, such as professional associations and unions, help their members maintain access to good jobs.
5. From Taylorism to Japanese production techniques, various management strategies are used to control workers and increase their productivity. Most strategies fall short on their claims to be participatory.
6. Research does not show a difference in men's and women's management styles. Differences are due to the position of managers or their organizational context.
7. Job satisfaction measures how workers feel about their jobs. Work and organizational characteristics are the primary predictors of how satisfied workers are.

8. Alienation is a structural condition of powerlessness that arises from the organization of work in the capitalist economy. Workers respond to alienating conditions in various ways, such as by engaging in sabotage or quitting their jobs. Strikes and other collective forms of resistance may have some success in changing the conditions of work.

Web Links

Job Futures

http://jobfutures.ca

Hilarious! Take the quiz on the home page of this website and see how many occupations will come up for you! I answered all of the questions honestly and literally, a squillion jobs came up—yes, one of them was university professor—so the test is likely a pretty good one!

Job Quality

http://www.jobquality.ca

Cool. This Canadian site measures how Canadians and job quality/satisfaction! Check it out for a number of reasons—insight to the professions that you want to enter, their research protocol, and how Canadian workers are compared to other countries.

Technology in the Workplace

http://www.abilities.ca

This site links people with disabilities to resources relating to education, employment, training, technology and more. A very interesting site for those who are unfamiliar with the disabled world, and an incredible resource for those who live with challenges.

The Work and Lifelong Learning Research Network (WALL)

http://www.wallnetwork.ca

Swing by this website and you can see what I'm doing for my doctoral work—I'm working with Dr. Peter H. Sawchuk and we're looking at Worker Learning and Technological Change. It's actually a pretty sweet gig! Being a part of WALL has allowed me study in all areas of work and lifelong learning.

Canadian Legal FAQs

http://www.law-faqs.org/nat/unions.htm

The site provides a list of frequently asked questions about various legal areas in Canada, and this link specifically takes you to an extensive history of unions in Canada as well as a list of the unions found in Canada.

Critical "Linking" Question

This one is sort of a no-brainer, but good practice nonetheless. Visit the Job Futures website and initially enter your information honestly to see what occupations come up.

Then, fiddle with the information that you enter, just to see what other kinds of occupations come up. How accurate do you think these sorts of tests are? As you can tell by my description of the site, it worked for me. Will it work for you too?

Solutions

True or False?

1. T
2. T
3. F
4. T
5. F
6. T
7. F
8. T
9. T
10. F

Multiple Choice

1. B
2. C
3. C
4. A
5. B
6. A
7. C
8. A
9. D
10. B
11. B
12. A
13. D
14. C
15. D

Chapter 12
Education

Chapter Introduction

Though my family calls me a "professional student," I prefer to refer to myself as a lifelong learner, and can't imagine my life without being involved in some form of education. I'm not exactly sure why I'm such an "academia nut," especially when no one else in my family is, but I do know that it has nothing to do with the money that I might earn in my career. Studies certainly show that since I'm from a lower-class background—I'm second-youngest of thirteen children—I'm not supposed to have achieved such high levels of educational attainment—in this chapter you'll see that this is often thought to be one of the "iron laws of sociology." I still can't believe that my dream of having a doctoral degree would ever come true, but by the time you read this, I'll likely just be getting ready to graduate! That's why I tell all of my students that you can only go as far as you dream. How much formal education will your job require? These should be the best years of your life—if you can find the perfect balance between going to school and having fun. I don't really have a personal philosophy but I've always believed that school should never have to hurt—if it does, then you're doing something wrong. Oftentimes that just means tweaking your schedule or cutting down back on the hours that you spend at a part-time job, but you should be able graduate with a smile on your face and be proud of your accomplishments while you look forward to using your skills in the workforce.

In this chapter, you'll explore how sociological theory and research can shed light on the operation of schools in a modern industrial society such as Canada. You'll learn about the critical importance of education both to individuals and to all of society and how Canadian schooling has been marked by three broad trends. Additionally, this chapter is full of results from some very interesting research based on the inequalities that exist in the education system, and you can see how the socialization process has been affected by school throughout the past. You'll learn a lot about why you're learning in the first place!

Jump Start Your Brain!

In this chapter you will learn that:

1. Sociologists study links between schools and society, focusing on inequality, socialization, and social organization. Various theories include structural functionalism, Marxism, human capital theory, feminism, credentialism, and institutional theory.
2. Canadian schooling has been marked by three broad trends. First, to fulfill a mandate to retain the vast bulk of students in secondary levels, schools have expanded. Second, this expansion has necessitated a greater variety of accommodations, as educators attempt to address a range of student abilities and exceptionalities. Third,

demands for postsecondary credentials are generating a simultaneous trend toward more intense forms of competition.

3. Canadian school-level attainment shows a mixed pattern of inequality, in that it varies by class, gender, and race.
4. Socialization through education has become less religious and now lacks the hard-edged prescriptive tone it had the past. Reflecting trends in pedagogical philosophy, today's schools have a more indulgent quality, with teachers seeking use less punitive tactics to elicit compliance from students.
5. Canadian schools have changed from wielding traditional authority to wielding legal-rational authority and in the process have become more bureaucratic. Today, schools are being pressured to become more "accountable" and "market-like."

Quiz Questions

True or False?

1. Early in the twentieth century, most people did not complete high school and all throughout Canadian history, higher education was deemed to be a waste for "common people." True or False
2. Contest mobility uses highly structured streaming to restrict access to higher education, while sponsored mobility promotes more competition within a unitary structure. True or False
3. Although functionalists believe that the best students are rewarded and often enter the higher-paying professions, they also admit that the very design of mass public schooling ensures that people who are born disadvantaged remain disadvantaged. True or False
4. Females now drop out of high school in fewer numbers, graduate more often, enter universities in greater numbers, and score higher on many standardized tests. In short, women have surpassed men on most measures of education attainment. True or False
5. Research is almost unanimous in its conclusion: schooling increases gaps along socioeconomic lines, while these gaps diminish in the summer when students are not in school. True or False
6. Marxists claim that schools impose a hidden curriculum, but one that promotes obedience to authority, not cheery, modern values, and contend that public education is structured to support capitalism by creating a disciplined wage labour force. True or False
7. Affluent children have many more advantages and fare better in school because of their home environments. This is considered to be one of the "iron laws" in the sociology of education. True or False
8. Just prior to World War II, governments across Canada began to see education as increasingly necessary for economic prosperity and the development of individual citizens. Consequently they gave public schools a new mandate: to retain as many youth for as many years as possible. True or False

9. A recent study suggests that when programs deregulate their fees, they become more likely to accept student who have highly educated parents and more likely to have students from poorer backgrounds who are eligible for bursaries and scholarships. True or False

10. The idea of the shopping-mall high school was introduced in the mid 1990s—when thoughts of the double cohort were just starting to develop—and it accurately depicted how schools were then striving to accommodate students. True or False

Multiple Choice

1. Sociologists examine three major ways that schools connect to society. Which of the following is *not* one of them?

   a. Schools shape society.

   b. Schools are a primary source of social control.

   c. Schools socialize people.

   d. Education is about social organization.

2. Higher education can be seen as stratified along two main dimensions. Which two?

   a. research opportunities and faculty teaching credentials

   b. private versus public funding

   c. admission requirements and financial consideration

   d. selectivity of institution and field of study

3. The best-known Canadian university rankings are compiled by *Maclean's* magazine. Although popular, this practice has been criticized. Which of the following is *not* one of the criticisms, as stated in your text?

   a. the use of questionable measures

   b. creating an artificial image of hierarchy among universities

   c. creating artificial competition among universities

   d. researcher bias

4. Males once had a virtual monopoly on higher education, but this imbalance began to change in the late _____________.

   a. 1940s

   b. 1950s

   c. 1960s

   d. 1970s

5. Racial and ethic patterns of educational attainment show a large variation among minority populations. Which of the following does *not* account for this variation?

   a. differences in socioeconomic status

b. traditional gender inequality

c. immigration policies

d. colonization

6. According to one view, since schools reward only some students and deem others to be academically unfit, they create disincentives for unsuccessful students. Low-ranked students suffer an inglorious status and are often labelled ____________.

a. "low achievers"

b. "underachievers"

c. "losers"

d. "mediocraties"

7. ____________ theory asserts that the school's role is primarily economic: to generate needed job skills and it also assumes that both individuals and governments invest their time and dollars in schools because they believe it will lead to financial prosperity.

A. Human capital

B. Cultural capital

C. Human potential

D. Human investment

8. ____________ do not want students to be merely instrumental toward their schooling but aim instead to nurture intrinsic forms of motivation, to engage students and have them work voluntarily.

a. Progressive education theorists

b. Feminists

c. Supporters of free schools

d. Functionalists

9. The tutoring industry has undergone a staggering transformation in the last 30 years, with the number of tutoring businesses growing between ____________ and ____________ in major Canadian cities.

a. 10%; 50%

b. 80%; 120%

c. 200%; 500%

d. 300%; 600%

10. The idea of the ____________ was introduced in the mid-1980s and it accurately depicted how schools were then striving to accommodate students.

a. free "progressive" school

b. Montessori school

c. private school

d. shopping-mall high school

11. As schooling becomes ever more central to society in the modern era, it loses its "magical" quality to command traditional deference. This is a prime example of Max Weber's notion of _____________.

a. rationalization

b. disenchantment

c. the reality myth

d. postmodern education

12. Progressive pedagogy—which virtually revolutionized the education system—is rooted in the ideas of _____________.

a. John Dewey

b. Paolo Friere

c. A. S. Neill

d. Max Weber

13. Feminist critics argue that explicit gender stereotyping of the late nineteenth century has only been masked and they support their argument by pointing out several indicators beyond standard measures of school attainment. Which of the following is *not* one of them?

a. They examine school staffing and note that despite some change, most positions of power remain in the hands of men.

b. They purport that school textbooks are loaded with sexist language and illustrations that depict most active characters as male and stories that depict women in nurturing occupations.

c. They believe that schools tolerate the conduct of male students that, in the adult world, would be considered sexual harassment.

d. They claim that when applicants are equal on all other points of admission, preference is given to male candidates.

14. The theory of multiple intelligences contends that the traditional notion of intelligence based on IQ testing is an overly narrow model of human potential. How many kinds of intelligence does Gardner propose there are?

a. four

b. five

c. eight

d. ten

15. In the early 1960s, about _____________ of university graduates were women; in some fields, such as forestry, no women graduated, and several others had only a handful of female graduates.

    a. 10%

    b. 15%

    c. 25%

    d. 30%

Critical Thinking

1. Canada has a relatively small national market for undergraduate credentials. Although degrees from top-ranked colleges in the United States can offer great opportunities there, few employers in Canada value the name of any one Canadian university over others. Do you agree or disagree with this?
2. Higher education is becoming increasingly valuable. Required high school grades—the prime currency for entrance to universities—have steadily risen over the past decade. Many universities are boosting their entrance standards and tuition fees, gaining repute by admitting top students but also by rejecting large numbers of qualified students. How do you feel about this? Should grades be the only thing that universities look at when considering an applicant for admission? Could or should they grant prospective students an interview? If so, logistically, how could they do this? Would you prefer this method even if it meant a fee increase in the application process?
3. Although higher numbers of women are going to university, they often enter different fields than their male counterparts. Table 12.3 on page 316 shows that almost 58 percent of all students in Canadian universities are female and that women form the majority in all fields except math, computer science, architecture, and engineering. Though for sure women have made great strides in representing themselves in this area, why do you think these fields continue to be male-dominated?
4. In the introduction to this chapter, I've told you that I'm hoping to graduate with my doctoral degree soon. What I didn't tell you—and will not tell you—is how long it's taken me to get it! I often laugh when I see students in my office who are so panicked about finishing school because they want to hurry up and get out into the working world—students much like yourself probably, who are less than 21 years old! Read Box 12.2 on page 324 and tell me if you're worried about how soon you'll finish your schooling. What's the big hurry? Do you not realize that you have the rest of your life to work? Relax—and enjoy the time that you have in school—working will come sooner than you know!

Time to Gear Down!

At the conclusion of this chapter, you should be able to discuss or write about the following, without having to rely on the textbook:

1. More of the selection function of school systems is being transferred from secondary to postsecondary levels. In past decades, secondary-level streaming played an important gatekeeping role, so an adolescent's life-chances were shaped largely by his or her performance in high school. However, as higher education expands, life-chances are determined increasingly by where a person graduates in a stratified structure of higher education.
2. As more Canadians attain advanced levels of education, formerly valuable credentials like the high-school diploma have been devalued. Devaluation generates demand for more schooling, leading to a spiral of educational expansion as groups jockey for advantages in the labour market.
3. A mixed pattern of inequality exists in Canadian school attainment. Along social class lines, students from more affluent backgrounds continue to enjoy considerable advantages. In terms of gender, women now outpace men on many indicators, though considerable gender segregation remains in certain fields of study. The evidence on racial and ethnic inequalities is more mixed. The legacy of conquest for Aboriginals and their history of discrimination and segregation continue to be manifest in schools, while changing immigration selection policies have created a partial reversal in patterns of attainment for other racial groups.
4. Today's moral education is less religious and explicit than before. Schools now mostly aim to have students understand key issues, rather than having a hard-edged, prescriptive tone. The hidden curriculum continues to emphasize the orderly completion of tasks, punctuality, and neatness. But today's schools have a more indulgent quality than they did in the past, trying to shore up students' self-esteem and to capture rather than command their interests.
5. Canadian schools have changed from informal institutions that wielded traditional authority to modern legal-rational bureaucracies. Within this organizational form, progressive educators have pursued strategies to spark intrinsic learning and cater to students in ever-more-accommodating ways. In the 1970s, this involved lessening teacher power in favour of students. Today, it sometimes involves the use of market-like mechanisms to treat students as if they were "customers."

Web Links

Maclean's

http://www.macleans.ca/education/index.jsp

While this link won't bring you to the famous—or infamous—ratings, it will bring you to the education section of *Maclean's* magazine, which you should read with a very critical mind—be aware of what and where your sources are from!!!

Scholarships

http://www.scholarshipscanada.com

Many students aren't even aware of the funding opportunities that might be available to them. Take the time to investigate and reconcile yourself to the fact that you might have to put up with some junk mail. The time that you invest into this might be very much worth it!

History of Education in Canada

http://fcis.oise.utoronto.ca/~daniel_sch/assignment1

This is an excellent reference site that offers brief descriptions of important events in educational history, organized by date, year, and decade. It's a work in progress and I go to school with the author—who's very nice!

Beyond Human Capital Theory

http://leo.oise.utoronto.ca/~dlivingstone/beyondhc

Caveat emptor! I will warn you ahead of time that this will indeed be a very tough read—*but* David Livingstone actually puts a new spin on one of my old favourite theories and argues that the human capital theory is not necessarily applicable in today's society of underemployment and overqualification.

Credentials Please?

http://www.belforduniversity.org/?engine=adwords!4580&keyword=%28degree+mills%29&match_type=

Why have I paid close to $40,000 for my doctoral degree when I could have bought one for only $549 from Belford University—one of my favourite online sites! Check this site out—one of many where you can buy any credential you like—and try to figure out how these institutions can operate.

Critical "Linking" Question

Visit Belford University and "shop" online for the degree or diploma of your choice. Degree mills are a very serious problem and it boggles my mind that these institutions flourish in the educational sector. Let's say that you actually had the nerve to buy a degree. When it came to doing a job that relied on your skills, how would you pull it off?

Solutions

True or False?

1. T
2. F
3. F
4. T
5. F
6. T
7. T
8. F
9. T
10. F

Multiple Choice

1. B
2. D

3. D
4. B
5. B
6. B
7. A
8. A
9. C
10. D
11. B
12. A
13. C
14. C
15. D

Chapter 13
Religion

Chapter Introduction

Did you know that for all intents and purposes, over 100,000 people have declared themselves members of the Jedi Knight religion in England, Australia, and Canada? Yup, it's true, and it's caused a bit of havoc in the statistical data for census takers in those countries. I suppose that it's quite likely the best example of just how sticky it can be when it comes to actually defining and measuring the concept of religion in our society today. When it seems as though every day brings a new story of heartbreak and devastation somewhere in the world, where do you go to find solace among the sadness? For many people, in different ways, religion is the pathway through which they find peace and comfort. Is this the case with you? Can you imagine pretending that you're Han Solo, crying on Chewbacca's shoulder? For some reason, that is just wrong on so many levels for me.

When you take the time to reflect on some of the harder times in your life, or, conversely, some of the celebrations in your life, what role—if any—has your religion played? How much time, if any, do you spend wondering about what happens when you die? The answers to these questions have kept world thinkers and philosophers busy for centuries, and I suspect that they'll be just as busy for years to come. Regardless of how you measure or determine your level of religiosity, I hope that you have some way or some one to help you through the tougher times in life as well as those times when you're especially happy, or proud of yourself—say graduation day? Most people say that life is too short, but I'm a pessimist and I always counter them by saying that life is too long—to be doing things that don't make you happy.

This chapter examines what some of the early social scientists had to say about religion and discusses how sociologists go about studying religion in both its individual and its group forms. Then you'll get a bird's-eye view of religiosity in Canada and some of the influences that religion has on both individuals and societies. The author ends this chapter with a reflection on the trends of religion that we might see in the future. As you read through this chapter, try to think about your own religious habits and compare them to the practices of your parents and friends. What are some of the things that influence your religious beliefs and how do your beliefs influence others in society?

Jump Start Your Brain!

In this chapter you will learn that:

1. Religion is something that can be examined by social scientists and has been studied since the beginnings of sociology.
2. Religion has both individual and social components—people display a wide range of levels of commitment, but groups play a major role in instilling and sustaining

personal religiosity. Since religious groups are organizations, they can best be understood by using organizational concepts and frameworks.

3. For the vast majority of people, religious commitment and involvement are rooted in social institutions, particularly the family.
4. Religion's influence on individuals tends to be noteworthy but not unique, while its broader role in most societies is to support social structure and culture.
5. Despite the numerical problems of some groups, religion's future in Canada and elsewhere is secure, grounded in ongoing spiritual interests and needs.

Quiz Questions

True or False?

1. Weber examined the possibility that the moral tone that characterizes capitalism in the Western world—the Protestant ethic—can be traced back to the influence of the Roman Reformation. True or False
2. According to Marx, those who hold power encourage religious belief among the masses as a subtle tool in the process of exploiting and subjugating them. True or False
3. Despite the problems of some groups, religion's future in Canada and elsewhere is secure, grounded in ongoing spiritual interests and needs. True or False
4. With industrialization and increased prosperity and stability, some of the smaller, independent evangelical groups evolved into sects. True or False
5. Full-time employment is associated with a noticeable decline in both attendance and the importance given to religion. True or False
6. Religion has been studied by sociologists since the 1960s due to the development of counterculture. True or False
7. For the vast majority of people, religious commitment and involvement are rooted in social institutions, particularly the family. True or False
8. The church-sect typology attempted to describe the central characteristics of these two types of organizations, as well as account for the origin and development of sects. True or False
9. When studying the membership of religious groups, it readily becomes apparent that the vast majority of those involved are following in parental footsteps. True or False
10. Weber argued that religion serves to hold in check the potentially explosive tensions of society. True or False

Multiple Choice

1. In _____________ view, religious rites provide guidelines pertaining to how people should act in the presence of the sacred.

a. Marx's

b. Durkheim's

c. Weber's

d. Comte's

2. Canadians exhibit relatively high levels of religious belief, practice, experience, and knowledge. Indeed, some ____________ in 10 Canadians say they believe in God.

   a. 6

   b. 7

   c. 9

   d. 8

3. In the 2001 census, ___________ of Canadians indicated that they have a religious preference.

   a. 84%

   b. 48%

   c. 78%

   d. 62%

4. After declining steadily for _______________, national weekly service attendance increased from 21 percent to 25 percent between 2000 and 2005.

   a. four decades

   b. three decades

   c. five decades

   d. six decades

5. The ________________ claims that religion—traditional or otherwise—persists in industrial and postindustrial or postmodern societies, continuing to address questions of meaning and purpose, and responding to widespread interest in spirituality.

   a. secularization thesis

   b. persistence thesis

   c. denominational thesis

   d. religiosity thesis

6. Durkheim argued that religion's origin is social. People who live in a community come to share common sentiments, and as a result a ____________ is formed.

   a. religion

   b. personal religiosity

c. charismatic leader

d. collective conscience

7. Sociologists can offer considerable insight into "the observable part" of religion. They can examine at least five areas. Which of the following is *not* one of the areas?

a. who tends to think they have experienced God

b. who believes in life after death and what individuals think will happen when they die

c. the financial cost to individuals in society taking part in a particular religion

d. the impact that religious involvement has on individuals and societies

8. Monotheism is related to goals of political unification and is a belief in _____________.

a. one church

b. one God

c. one religion

d. one denomination

9. Historically, Canada's "Bible Belt" has been viewed as _____________ when by every conceivable measure it probably has actually been the Atlantic region.

a. Saskatchewan

b. Ontario

c. Quebec

d. Alberta

10. Dimensions of religiosity are the various facets of religious commitment; Glock and Stark (1965), for example, identify four. Which of the following is *not* one of the facets?

a. belief

b. experience

c. faith

d. practice

11. The secularization thesis is the dominant explanatory framework that the media and _____________ use in making sense of religious developments in Canada.

a. sociologists

b. Statistics Canada

c. dioceses

d. Ministers

12. Which of the following age groups have a higher level of religious participation and commitment?

    a. age 11–17

    b. age 1–10

    c. age 35–54

    d. age 18–34

13. The rise of families with two parents in the paid labour force and the implications of the rise of this family form for the highly selective use of time may be largely responsible for the decline in mainline Protestant and Catholic attendance in Canada after the _______________.

    a. 1940s

    b. 1950s

    c. 1960s

    d. 1970s

14. During the 1980s and 1990s, _______________ contributed most of the growth to Hindu, Sikh, Muslim, and Buddhist groups in Canada.

    a. youth

    b. Canadian-born baby boomers

    c. Western Canadians

    d. immigrants

15. To the extent that groups can locate and respond to their "affiliates"—as it appears they have been doing in recent years—there is reason to believe that the embryonic "renaissance of religion" in Canada will _______________.

    a. remain constant

    b. end

    c. continue

    d. decline

Critical Thinking

1. As you've probably already gathered, much of the difficulty in conducting sociological research lies with the definition and measurement of the concept that you're trying to measure, and religion is no exception. If you were trying to determine levels of religiosity among Canadian adults, how would you define and measure religion? Would you change your definition if you were studying your classmates? If your study happened to fall close to a religious holiday, such as Easter or Diwali would your definition change? Why or why not?

2. The world's attention has been fixed on suicide bombing since 2001. Because such attacks are often religiously inspired, does it cause you to become more

aware of other religions? Does it make you want to learn more about those religions that the suicide bombers espouse? Are you in any way more suspicious of certain religions than others—specifically because of the events of September 11, 2001? If you feel as though you are more judgmental than perhaps you should be, how might you deal with these feelings?

3. In this chapter, the author states that religion is not going to disappear. Proponents of the secularization thesis expected religion to be replaced by science and reason as societies evolved, but opponents of this argument say that people will always need religion, in order to come to grips with the certainty of death. We certainly know that religion has certainly been reshaped and reformed over the years and it has certainly been redefined. What do you think? Is religion going to fall by the wayside? If so, what will replace it?

4. Well? Have you read *The Da Vinci Code* or seen the movie? Could you understand was all the fuss and bother was about? I did indeed read the book, but I'm not much of moviegoer. Admittedly, I was quite impressed by the knowledge that Dan Brown imparted, but not at all impressed with his writing skills. To what do you owe the level of success this book has reached? Is it merely because it has maligned the sanctity of the Church?

Time to Gear Down!

At the conclusion of this chapter, you should be able to discuss or write about the following, without having to rely on the textbook:

1. Sociology uses the scientific method to study religion, in contrast to religion, which explores reality beyond what can be known empirically.

2. The sociology of religion has been strongly influenced by the theoretical contributions of Marx, who stressed the compensatory role of religion in the face of economic deprivation; Durkheim, who emphasized both the social origin of religion and its important social cohesive function; and Weber, who gave considerable attention to the relationship between ideas and behaviour.

3. Religion can be defined as a system of meaning with a supernatural referent used to interpret the world. Humanist perspectives make no such use of the supernatural realm, attempting instead to make life meaningful.

4. Personal religious commitment increasingly has come to be seen as having many facets or dimensions. Four such dimensions are commonly noted: belief, practice, experience, and knowledge. Personal commitment is created and sustained by collective religiosity. In Canada, organized religion has experienced a considerable decline in participation in recent years, a trend that has had critical implications for commitment at the individual level.

5. The variations in the levels of individual commitment that characterize complex societies have led to explanations that emphasize individual and structural factors. Reflection, socialization, and deprivation have been prominent among the individual explanations, while the dominant structural assertion has been the secularization thesis.

6. At the individual level, religion appears to be, at best, one of many paths leading to valued characteristics, such as personal happiness and compassion. Although religion can be socially disruptive, Canada's emphasis on social cultural diversity functions to put limits on how religion can be expressed, thereby optimizing the possibility of religions contributing positively to social and collective life.
7. Although proponents of secularization saw religion as being replaced by science and reason, it now is apparent that religion continues to be important throughout the world, including Canada. Its future is not in doubt.
8. The search for alleged religious switchers and dropouts in Canada reveals that few have turned elsewhere or permanently opted for "no religion." Most still identify with the country's established groups.
9. Canadians young and old, in very large numbers, continue to hold religious beliefs, claim religious experiences, and express spiritual needs. Many also say they are receptive to greater involvement with religious groups.
10. To the extent that groups can locate and respond to their "affiliates"—as it appears they have been doing in recent years—there is good reason to believe that the embryonic "renaissance of religion" in Canada will continue.

Web Links

The Rick Ross Institute for the Investigation of Cults

http://www.rickross.com

OK, OK, I couldn't resist. Skedaddle to this site and see what it's like to hire a real live Cult Investigator. Trust me, there is some very wild stuff going down here.

Liberty Magazine

http://www.libertymagazine.org

Simply marvellous! This free online magazine pulls out all the stops when it comes to discussing religion—of any kind and the social issues of the day. Extremely hilarious articles are mixed with insightful, unbiased, and incredibly helpful links. A real gem of a find.

Civil Religion

http://www.facsnet.org/issues/faith/sherrill_indy.php

Civil religion is often linked to the United States, and as soon as you hop on this link you'll know why it's as American as apple pie! I'm wondering if you think that mixing religion and politics is a good idea—does it seem obvious to you here?

Da Vinci Debunked!

http://altreligion.about.com/library/bl_davincicode.htm

Go on—you know you want to! Read all the hoopla that's got the Catholic Church in an uproar!

Ontario Consultants on Religious Tolerance

http://www.religioustolerance.org

Scoop of the century! This extensive site has squillions of links to information on various religions, but the main focus is to promote religious tolerance, regardless of peoples' beliefs. Wickedly awesome—again, no pun intended!

Critical "Linking" Question

Jump on the *Liberty* magazine site and click on "Legal Resources." When you're there, check out the Canadian Resources section and browse through some of the issues that are happening in Canada. Do you get a different feel from the coverage of those issues than you do from the rest of the articles in the magazine? Pay special attention to the topics that deal with homosexuality.

Solutions

True or False?

1. F
2. T
3. T
4. F
5. T
6. F
7. T
8. T
9. T
10. F

Multiple Choice

1. B
2. D
3. A
4. A
5. B
6. D
7. C
8. B
9. D
10. C
11. B
12. D
13. B
14. D
15. C

Chapter 14
Deviance and Crime

Chapter Introduction

Well, well, well. Things have really changed since I was an undergraduate student many years ago. I would have never thought of going to class with blue hair or a pierced eyebrow—although I was the first girl in my high school to have two piercings in my left ear, and let me tell you, it caused quite a stir back then! I even got in big trouble at home. It was a very, very, long time ago and teenage girls really didn't get their ears pierced, let alone get two piercings in one ear. It was "just weird, ya know?" Now that I'm teaching undergraduate students, those things that I once thought were wildly deviant are just a normal part of everyday life in college. In fact, my measly two piercings are nothing compared to what I see on a daily basis at school. I can't count the number of times I've asked a student about the "ouch factor" on a particular piercing or tattoo—in other words, how much it hurt to get done. Deviance is one of my favourite topics to teach, because everyone has such a different notion of what is and what is not deviant. Many times when people think of someone who is deviant, they think of a criminal or someone who has done something horrible. Often we don't realize that deviance has many forms—an incredibly gifted child, an extraordinary athlete, or a man wearing a kilt are all examples of how we define deviance in society—and if you think about it, you realize that deviants are not always the bad guys! Deviance occurs in all societies but is not necessarily defined or socially constructed the same in each culture. For example, a man wearing a kilt may not be seen as deviant in Scotland, but he may get a couple of odd looks in Canada. When I think of deviance, I always remember the adage "One man's junk is another man's treasure." Deviance, because it is a social construction, can take on different forms and different meanings in all societies. More often than not, though, most people, when they think of deviance, think of crime, and this chapter really focuses on that as well.

This chapter discusses the different ways in which sociologists have thought about deviance. You'll learn that some are interested primarily in where rules about deviance come from, how and why these rules change, and the consequences of labelling people or behaviours as deviant. You'll also see that other theorists focus on why some people tend to become rule-breakers. Making that important connection between theory and reality, the final part of this chapter looks at recent trends in Canada as well as changes in crime and social control around the world. A central theme of this chapter is that deviance is not simply about marginal people and odd behaviours. Studying how deviance is defined and how people react to it tells us about how a society is organized: how power, privilege, and resources are distributed and how social order is achieved. As you are probably already aware, there's a lot more to deviance than meets the eye!

Jump Start Your Brain!

In this chapter you will learn that:

1. What becomes defined as deviant or criminal depends on the circumstances of time and place because definitions of deviance and crime are subject to conflict and change. These definitions are the outcome of political processes and power relations in which different groups compete to define right and wrong.
2. There are many consequences of defining deviance. Reacting to deviance can increase a group's social solidarity and clarify its moral boundaries. But these reactions are also an important source of deviance and may lead to more serious and organized forms of deviance.
3. Sociological explanations of deviance can be grouped into two types: those that emphasize the social factors that motivate or allow people to engage in deviance and those that emphasize the political processes and power relations that result in some individuals and behaviours being defined and treated as deviant while others are not.
4. Crime rates have been declining in Canada during the past decade; however, rates of serious crime are still higher in Canada than in many other industrialized countrics and arc distributed unequally among various groups.
5. Legal reactions to crime in Canada balance two goals: protecting society and protecting the rights of those accused of crime. Traditionally, the protection of society has received greater emphasis in the legal process. Canada incarcerates more people per capita than most industrialized countries but is increasingly relying on alternatives to incarceration for less serious criminal offences.

Quiz Questions

True or False?

1. What becomes defined as deviant or criminal depends on the circumstances of time and place because definitions of deviance and crime are subject to conflict and change. True or False
2. Crime rates have been increasing in Canada during the past decade. True or False
3. Formal social control is practised by the state through official organizations and agents primarily outside of the criminal justice system. True or False
4. A potential benefit of deviance, according to Marx, is the clarification of a group's moral boundaries. In defining and confronting deviance, a group highlights its standards of right and wrong to its members. True or False
5. Opportunity theories do not try to explain why people decide to commit crime. They simply assume that many people lack the controls to stop themselves. True or False

6. Contrary to what many Canadians believe, the crime rate in 2004 was about the same level as in 1979, and the homicide rate was 36 percent lower than in 1975. True or False
7. Although Canadians do not appear to be at greater risk of criminal victimization now than they were in the 1970s or 1980s, they continue to be at greater risk than citizens of some other industrialized nations. True or False
8. Some learning theories argue that long exposure to violent images desensitizes young people to violence and conveys the message that violence is an acceptable way to respond to frustration. True or False
9. In Canada, crime rates have historically been lower in the territories and highest in the Atlantic provinces. True or False
10. The crime-control model is a model of criminal process that emphasizes reducing crime and protecting society by granting "deputy" officials broad powers. True or False

Multiple Choice

1. Although deviance is defined by negative social reactions, it can have positive consequences for some social groups. This notion was an important theme of ____________ work.
   a. Emile Durkheim's
   b. Karl Marx's
   c. Robert Merton's
   d. Max Weber's
2. According to the ____________ perspective, the various social, economic, political, ethnic, religious, and professional groups that characterize modern societies continually compete with one another for status and influence.
   a. institutional competition
   b. social conflict
   c. status-conflict
   d. laissez-faire
3. The lack of fit between cultural goals and social structural opportunities, called ____________, means that many people feel strained because they are not able to achieve what they have been taught to value or desire.
   a. differential association
   b. alienation
   c. anomie
   d. strain theory

4. One example of a(n) ____________ is the campaign by Mothers Against Drunk Driving (MADD) to educate the public about the costs of drunk driving, to stigmatize those who drink and drive, and to increase criminal penalties for drunk driving.

   a. crime-control model

   b. moral crusade

   c. due-process model

   d. ethical campaign

5. An important contribution of critical perspectives on deviance is their focus on the consequences for the people and the groups who are targets of the deviance-defining process. The ____________ perspective is organized around the idea that societal reactions to deviance are an important cause of the deviance.

   a. labelling

   b. strain theory

   c. learning theory

   d. differential association

6. _______________ is affecting all arenas of human behaviour, including criminal behaviour.

   a. Rural living

   b. Unemployment

   c. Social class

   d. Globalization

7. There are at least three ways in which learning theories can be applied to explain deviant behaviour. Which of the following is *not* one of them?

   a. Long exposure to violent images desensitizes young people to violence and conveys the message that violence is an acceptable way to respond to frustration.

   b. Some individuals, regardless of their social status or gender, are genetically predisposed to engaging in criminal behaviour.

   c. Many types of crime require opportunities to learn specific techniques and procedures.

   d. Many crimes are simply the imitation or modelling of others' behaviour.

8. Crime committed by legitimate business organizations includes violations of antitrust, environmental, food and drug, tax, health and safety, and corruption laws and is called corporate crime. The economic costs of these crimes have been estimated at ____________ times those associated with street crimes.

   a. 20

   b. 30
   c. 40
   d. 50

9. Males are currently _____________ as likely as females to be victims of homicide.
   a. three times
   b. twice
   c. six times
   d. four times

10. The *Youth Criminal Justice Act* allows an adult sentence for any youth aged _____________ and older convicted of a serious crime, increases the scope of cases for which youths can presumptively receive an adult sentence, lowers the age at which youths found guilty of certain offences will presumptively be sentenced as adults, and permits judges to consider allowing the publication of names of some youths found guilty of serious violent offences even if they were not given the adult sentence.
    a. 13
    b. 14
    c. 15
    d. 16

11. _____________ incarcerates more people per capita than most industrialized countries, but it is increasingly relying on alternatives to incarceration for less serious criminal offences.
    a. The United States
    b. Canada
    c. Sweden
    d. Denmark

12. _____________ is based on a social rather than a legal view that criminal acts not only injure a victim but also affect communities and offenders and that therefore solutions need to involve all three parties.
    a. Differential association
    b. Moral crusade
    c. Restitution
    d. Restorative justice

13. The public is becoming more aware of the high cost of imprisonment. Canada spends more than $2 billion on adult and youth corrections. For example, in

2003–04, it cost ____________ per day to keep one person in federal prison in Canada.

a. $250.08

b. $240.18

c. $280.40

d. $236.18

14. ____________ people are also overrepresented among those serving time in Canada's jails and prisons.

a. Aboriginal and Asian

b. Youth and Aboriginal

c. Black and Aboriginal

d. Male and Aboriginal

15. For ____________ theorists, conflict between different classes, and the desire of the capitalist class to control the working class, results in some behaviours (such as burglary) being targeted for control more than other behaviours (such as corporate crime).

a. functional

b. strain

c. Marxist

d. symbolic interaction

Critical Thinking

1. Let's say that you decide to go into sociology as a profession and that you want to study social deviance—maybe even become a forensic profiler or something very cool like that. Which of the two sets of theories would you focus on—motivational theories that ask why people commit crimes, or control and opportunity theories that ask why doesn't everyone commit crimes? Check out these theories again in your text before you answer. Admittedly, my favourite is Merton's strain theory.
2. In 2006, Robert Latimer, who has served half of a ten-year sentence for killing his daughter in 1993, says he hopes the new federal government and Canada's top court will intervene in his case. Though he is eligible for parole on December 8, 2007, Latimer hopes that Stephen Harper will intervene on his behalf and bring his story into new light. Though it's doubtful that the mercy killing debate will ever end, read Box 14.1 on page 368 and decide how you feel about what Robert Latimer did.
3. The success of the MADD campaign reflects the group's ability to gain the support of the media and various politicians, through which both legal and popular definitions of drunk driving were changed. Do you think that groups such

as this one would be as successful if they didn't have the support of the media? Do you think that by having a "designated driver" it condones the behaviour of those who aren't driving the car? For example, have you ever said, "Well, I could get totally wasted, because I'm not driving."

4. In 2002 a Superior Court found that Mark Hall's rights to freedom from discrimination on the basis of sexual orientation were being violated when he won the right to take his boyfriend to the school prom at his Catholic high school. Since we know that much of deviance is dependent upon the reactions of others, how would you have felt if this occurred at your high school? Would you have supported Mark Hall's right to bring his partner or the Catholic School Board's ban?

Time to Gear Down!

At the conclusion of this chapter, you should be able to discuss or write about the following, without having to rely on the textbook:

1. Deviance and crime are defined by the social reactions to them. Both what is defined as deviance and the way people react to deviance depend on social circumstances. Even what you may consider to be serious deviance is subject to conflicting opinions and changing social reactions.

2. Crime is a special case of deviance and is defined by social norms that are formalized in criminal law. The response to crime is through formal social control—such as the criminal justice and correctional systems—whereas the response to noncriminal deviance is through informal social control—such as gossip, avoidance, and other forms of disapproval. There tends to be wide—though certainly not complete—agreement that crime is wrong but much less agreement that noncriminal forms of deviance are wrong.

3. Deviance is defined through a political process that typically involves struggles between competing groups over status, resources, knowledge, and power. Although some relationship often exists between the harm that a behaviour causes and the likelihood that the behaviour is defined as deviance, many harmful behaviours are not defined as deviant and some behaviours defined as deviant are not harmful.

4. Sociological explanations of deviant behaviour can be grouped into two types: those that emphasize the social factors that motivate or allow people to engage in deviance, and those that emphasize the political processes and power relations that result in some people and behaviours being defined and treated as deviant while others are not.

5. Rates of crimes have decreased in Canada in recent years. Canadians are less likely to be victims of serious violent crime than people in the United States, but Canadians face higher risks of homicide than citizens of other developed democracies. Aboriginal Canadians are at particularly high risk of involvement in criminal homicide—both as victims and as offenders. They are also

overrepresented among those incarcerated in Canadian prisons and jails. This has been attributed in part to systemic racism in the criminal justice system.

6. Crime rates are often affected by major social changes. For example, rapidly changing technologies have created new opportunities for both personal and property crime and the globalization of the world's economy has been accompanied by the globalization of corporate and organized crime. Changes in gender inequalities, however, have not reduced the gender gap in criminal offences. Men still greatly outnumber women among offenders.
7. Canada has traditionally favoured the crime-control model of the criminal process and, consistently with that model, relies heavily on incarceration. More recently, however, this emphasis has shifted. With the introduction of the Charter of Rights and Freedoms, the due-process model has gained ground. And although Canada continues to incarcerate offenders at higher rates than most other Western democracies, Canadian incarceration rates have declined in recent years as the criminal justice system explores alternatives to imprisonment.

Web Links

Department of Justice Canada

http://justice.gc.ca/en/jl/index.html#statistics

Use this site to review the justice system in Canada and all of its departments, levels and provincial differences. There's also a smokin' interactive link that prepares you if you have to make a court appearance!

MADD Canada

http://www.madd.ca

This is the home site for Mothers Against Drunk Drivers—an excellent example of a moral crusade and their power with justice in Canada.

International Centre for Criminal Law Reform and Criminal Justice Policy

http://www.icclr.law.ubc.ca

Hey—just a wickedly decent site with some very solid Canadian research happening here. An international agency affiliated with the United Nations this centre is located in British Columbia and assists with implementing criminal justice policies on a global level.

The Centre for Restorative Justice

http://www.sfu.ca/crj

Scoop! Further to the textbook discussion, this link will provide you with an in-depth look at restorative justice in Canada. Very cool and highly intriguing. What do you think of all this?

crimeinfo

http://www.crimeinfo.org.uk/index.jsp

Pip Pip and Cherrio! OK—that was really bad, but this site is *not.* It's a ab-fab site on crime but it's based in the United Kingdom. That totally doesn't mean that you won't learn anything—it means that you'll learn about deviance and crime from a different perspective and this site is just so flippin' cool!

Critical "Linking" Question

A tough decision here, but I'm going to go with restorative justice. Take a walk on the wild side and visit the Centre for Restorative Justice. I was especially interested to read the papers that had been done on Aboriginal peoples. How successful do you think these programs will be?

Solutions

True or False?

1. T
2. F
3. F
4. F
5. T
6. T
7. T
8. T
9. F
10. F

Multiple Choice

1. A
2. C
3. C
4. B
5. A
6. D
7. B
8. D
9. B
10. B
11. B
12. D
13. B
14. C
15. A

Chapter 15
Population and Urbanization

Chapter Introduction

I remember my very first trip to Toronto. I was about 12 years old and I can still remember the overwhelming feeling that I got when I saw a homeless man for the first time in my life. It was devastating for me and more than a bit confusing. I had no idea that homeless people even existed. I remember giving him all the money that I had to spend (which wasn't much) and walking away feeling totally inadequate. We didn't have homeless people in the city where I grew up, so seeing the poverty that existed in a city that wasn't so far away from home really opened my eyes to what was going on in the world. On the other hand, I also remember how tiny I felt standing next to such tall buildings and actually being afraid for the people who had to work and live on the top floors—I guess I must have had some sort of elevator/escalator thing happening, because I'm still a little skittish on escalators and those revolving doors really freak me out. I quickly became more aware of the differences between urban and rural life, and more importantly what I saw in my mind, the differences between rich and poor. For a while I tended to blame homelessness on the city planners for not having the foresight to see that what they were building was not really affordable housing for a large segment of the population. I've since learned that homelessness is about much more than the physical structures that define a society and I now know that it can be explained by the social institutions and relations that exist in our world. Just knowing about this is hardly enough to eliminate it, but I think that the greatest place to initiate change is in our minds and hearts. I think that, by now, you should have a greater understanding of how difficult it is to make any changes in society, and the bigger the issue that you want to change, the more resources you need to make it happen.

In this chapter you'll learn about the dual influence of environment and structure on city life over the course of the twentieth century. You'll also come to understand that there are three main types of cities: the industrial city, the corporate city, and the postmodern city.

Jump Start Your Brain!

In this chapter you will learn that:

1. The choice of opportunities, experiences, and lifestyles available to urban residents are shaped and constrained by three sets of factors: those related to the physical environment; those associated with population size, distribution, and composition; and those dictated by the changing structure of the wider economy.
2. Contrary to the theory that the core of Canadian cities is hollowing out like the hole in a doughnut, recent evidence suggests that city centres are remaining relatively stable economically and demographically; in contrast, the ring of older

suburbs around the city core is stagnating, losing jobs and population to a constellation of edge cities in the surrounding region.

3. The fastest-growing middle-class neighbourhoods today are private communities, where control over local government, public services, and security rests in the hands of nonelected, professional managers; one popular form of such neighbourhoods is the gated community, where nonresidents are considered intruders and are monitored and kept out by guards, alarm systems, and surveillance cameras.

4. Cities are becoming multiethnic, with many neighbourhoods in the largest cities becoming home to members of visible minorities. This is generating racial and ethnic residential and commercial patterns that stand in marked contrast to the ecology of the industrial city.

Quiz Questions

True or False?

1. Cities are relatively large, dense, temporary settlements in which the majority of the residents do not produce their own food. True or False

2. Gentrification is the transformation of working-class housing into fashionable downtown neighbourhoods by middle- and upper-income newcomers. True or False

3. Malthus argued that while food supplies increase slowly, populations grow quickly. Because of this presumed natural law, only war, pestilence, and famine can keep human population growth in check. True or False

4. The fortress city is a city obsessed with urban security. True or False

5. According to Statistics Canada's most recent population projections, by the year 2017 nearly 1 Canadian in 3 is likely to be a member of a visible minority group. True or False

6. The tremendous growth of technology, such as fax machines, cellular phones, and e-mail, has geographically freed many employees whose linkages to the workplace are now activated on the road or from home offices. True or False

7. The fastest-growing upper-class neighbourhoods are private communities, where control over local government, public services, and security rests in the hands of non-elected, professional managers. True or False

8. For much of human history, societies hovered in a steady state in which both birth rates and death rates were low. True or False

9. Over the last decade of the twentieth century, Toronto and Vancouver (together with Calgary) made a major comeback as growth centres. True or False

10. The dual city has no dominant single core or definable set of boundaries. True or False

Multiple Choice

1. The choice of opportunities, experiences, and lifestyles available to urban residents is shaped and constrained by three sets of factors. Which of the following is *not* one of them?

   a. those related to the physical environment

   b. those related to the political climate

   c. those associated with population size, distribution, and composition

   d. those dictated by the changing structure of the wider economy

2. The ____________ posits that people actively choose where they want to live, depending on the extent to which a particular place either meshes with or constrains their preferred lifestyle.

   a. demographic transition

   b. concentric-zone model

   c. suburbanism

   d. environmental-opportunity theory

3. With just fewer than ____________ people inhabiting 9 million square kilometres of land, Canada remains one of the most sparsely populated nations.

   a. 32 million

   b. 10 million

   c. 18 million

   d. 54 million

4. The edge city is made up of three overlapping types of socioeconomic networks. Which of the following is *not* one of them?

   a. household networks

   b. networks of consumption

   c. networks of production

   d. networks of communication

5. In the early 1950s, ____________, Canada's first mass suburb, was built on the northern fringe of Toronto.

   a. Scarborough

   b. Don Mills

   c. Mississauga

   d. York

6. Marx insisted that ____________ is organized to keep the working class in a perpetual state of poverty and unemployment, criticizing Malthus' theories.

a. capitalism
b. communism
c. socialism
d. collectivism

7. Burgess' ____________ conceptualized the expansion of cities as a succession of concentric rings, each of which contained a distinct resident population and type of land use.
   a. concentric-nuclci model
   b. concentric-sector model
   c. concentric-zone model
   d. concentric-urban model
8. Some countries in Asia, Africa, and Latin America have exhibited distinctive patterns of ____________ whereby the rural and the urban have become blurred in unplanned settlements on the outskirts of large cities.
   a. suburbanism
   b. urbanism
   c. semi-urbanism
   d. peri-urbanization
9. A popular form of a private community is the ____________, where non-residents are considered intruders and are monitored and kept out by guards, alarm systems, and surveillance cameras.
   a. gated community
   b. postmodern city
   c. industrial city
   d. corporate city
10. There are four aspects to the postmodern city, which of the following is *not* one of them?
    a. The postmodern city is the product of the simultaneous operation of the forces of re-urbanization and counter-urbanization.
    b. Cities are becoming multiethnic, with many ethnic neighbourhoods in the largest cities becoming home to visible minority groups.
    c. The postmodern city is characteristically fragmented, even chaotic.
    d. The postmodern city is characterized by attempting to eliminate the privatization of public space.

11. Three theories have been proposed to explain the relationship between suburban residence and lifestyle patterns. They are _____________, selective migration, and class and life-cycle stage.

    a. functional

    b. structural

    c. gentrification

    d. suburbansion

12. For much of human history, societies hovered in a steady state in which both birth rates and death rates were very high. Women in preindustrial societies were destined to bear a large number of children. Which of the following is *not* one of the reasons why?

    a. Infant and mortality rates were very high.

    b. Children were expected to help with farm work.

    c. The human life span was much shorter.

    d. Contraception was expensive and only available to the upper classes.

13. A key feature of the suburban lifestyle is its emphasis on _____________.

    a. economy and politics

    b. income and social class

    c. children and the family

    d. surveillance and security

14. Today, nearly _____________ of the world's urban population resides in the less developed regions of Asia, Oceania, Africa, Latin America, and the Caribbean.

    a. 3/4

    b. 1/2

    c. 2/3

    d. 1/4

15. Louis Wirth proposed that the city is characterized by the concurrent trends of increasing size, density, and heterogeneity. In his view, the city creates a distinct way of life—"urbanism"—that is economically efficient but socially destructive. Which of the following does *not* appear on Wirth's short list of urban characteristics?

    a. decline of the family

    b. the disappearance of the old neighbourhood

    c. the undermining of traditional bases of social solidarity

    d. an increasing reliance on technology for communication purposes

Critical Thinking

1. I was doing my course work for my doctorate at OISE/UT during the whole "Tent City" fiasco in the summer of 2002. It was in the news everywhere and a highly contentious issue. In fact, I spoke with a man who had a full-time job but because rent was so high in Toronto, he lived in Tent City. I was absolutely shocked to learn about this. The dismantling of his "home" was devastating to him and so many more. What are your thoughts on the Tent City Eviction?
2. The fortress city, in which the urban disadvantaged are isolated socially and spatially from office workers, tourists, and suburban day-trippers sounds like an extreme example of city living, but this type of community is becoming more and more popular. These fortress cities remind me very much of the situation that I found myself in a few years back. I decided to treat myself to a much-needed vacation and went on one of those "all-inclusive" packages to the Dominican Republic. As long as I stayed on the highly guarded compound, I felt like I was living in paradise, but as soon as I walked outside the compound gates into the neighbouring village, it was a very different world. I felt somewhat guilty enjoying myself after seeing the poverty just beyond the gate and I came back from my vacation more than a little unsettled. Have you ever found yourself in this situation or a similar one? What were your feelings?
3. Before World War II, North American cities such as Toronto were configured in a grid, with residential avenues crossing long commercial streets at right angles. This spawned a lively "front-yard culture" in which passersby regularly interacted with front porch sitters, since front yards and families faced the street rather than the house itself. I remember walking to the corner store with my mother when I was young and distinctly recall that the trip could sometimes take an hour or so, when the store was only five minutes away. We would stop and chat with all the neighbours that were out doing yard work, or just enjoying the summer sun. Is a trip to your corner store anything like that? Do you know the names of everyone that lives on your street? Do you think that it has more to do with the actual placement of the houses or more to do with changing values in today's society?
4. OK—now you should know by now that I don't watch TV, but my students have told me about a reality show that was on a while back that was called *Amish in the City*. Apparently, the plot revolved around five Amish teenagers experiencing "English" (non-Amish) culture by living in a house with six mainstream American teenagers. Somehow, this just didn't sit very well with me. In this question, I don't want you to think about the TV part of things, but rather how well you think those teenagers would adapt to urban life had they not had the chance to do so through that show. Could you switch your lifestyle from an urban to a rural one very easily?

Time to Gear Down!

At the conclusion of this chapter, you should be able to discuss or write about the following, without having to rely on the textbook:

1. Cities are defined as relatively large, dense, permanent settlements in which the majority of the residents do not produce their own food.
2. Until the Industrial Revolution, cities were incapable of supporting more than about 5 percent of the total societal population, largely because of the absence of agricultural surpluses large enough to feed a huge urban population. In addition, high mortality rates, especially among children under ten years of age, dictated that, even with high birth rates, urban population growth was limited.
3. The best-known urban-growth model in the social sciences is Ernest Burgess' concentric-zone scheme in which the expansion of cities is conceptualized as a successive series of rings and circles, each of which segregates a distinct resident population and type of land use. Other, more recent explanations favour patterns of urban growth resembling pie-shaped wedges that develop along transportation routes or multiple nodes of economic activity each with its own nucleus.
4. In contrast to the rise of the city in Western Europe and North America, cities of the South have grown at a much faster rate than the industrial economy. The resulting "overurbanization" has accelerated problems of poverty and unemployment, which are rooted in basic structural inequalities and uneven development.
5. The corporate city of the 1950s and 1960s was the product of an urban-growth machine in which a coalition of politicians, planners, real-estate developers, business people, and other interest groups joined forces to engineer economic development and progress. The main products of this alliance were (a) the corporate suburbs, (b) shopping centres, (c) suburban industrial parks, (d) downtown office towers, and (e) high-rise apartment buildings.
6. Three alternative theories—structural, selective migration, and class and life-cycle stage—have been proposed to explain the relationship between suburban residence and lifestyle patterns. Although all these theories have merit, the suburban way of life observed by many researchers in the 1950s and 1960s appears to have been a unique product of a particular time and place.
7. In recent decades, the contemporary city has mirrored and incorporated a bifurcated global economy. On the one hand, the members of the upper-tier informational city work in jobs related to financial services, telecommunications, and high technology, and live either in gentrified downtown neighbourhoods or in private communities on the edge of the city. On the other hand, the members of the informal economy are excluded from the informational city and live in ethnic or racial ghettos that are often, but not necessarily, located in the inner city. Together, the informational city and the informal economy constitute the dual city whose residents have little in common with each other.
8. Cities are becoming multiethnic, with many ethnic neighbourhoods in the largest cities becoming home to visible minority groups. This is generating racial and ethnic residential and commercial patterns that stand in marked contrast to the ecology of the industrial city.

9. One of the defining characteristics of the postmodern city is the increasing privatization of public spaces. This is manifested in the booming growth of gated communities and other private residential enclaves where outsiders are not welcome and in the construction of fortress cities where tourists and other affluent consumers are kept in while the homeless and the urban poor are kept out.

Web Links

Massive Change

http://www.massivechange.com/exhurb

This site takes both a celebratory and cautious look at global capacity for continued urbanization. Check out the three questions that you have to answer when you submit your résumé!

Canada's Urbanization

http://www.sustreport.org/home.html

The Sustainability Reporting Program is Canada's first independent initiative to find out how we are doing at living in balance for the long term. Investigate this site to see what sustainability is all about.

DestinEducation

http://www.destineducation.ca/intro_e.htm

This is a very cool and highly informative site for anyone who might be interested in either studying or working in another country. This site also offers lots of tips for coping with change.

World Urban Forum

http://www.wuf3-fum3.ca/en/home.shtml

World Urban Forum, established by the United Nations, examines rapid urbanization and its affect on increased poverty, pollution and existing global infrastructures. Canada hosted the Forum in 2006 and you can click on the webcasting links to see the plenary sessions and lectures that were held during that conference.

Amish Life

http://www.amish.net/lifestyle.asp

OK—I know I'm totally being stereotypical here and I do not want to offend anyone, *but* when I think of rural, I always think of the Amish. I know, I know—completely insane! Check out this site though—it's awesome to compare urban and rural life!

Critical "Linking" Question

As if you didn't know! Take a tour of Amish country and really get a sense of what a different culture is all about. Are there values and beliefs different than ours, or is it just the tangible things that make this way of life so different?

Solutions

True or False?

1. F
2. T
3. T
4. T
5. F
6. T
7. F
8. F
9. T
10. F

Multiple Choice

1. B
2. D
3. A
4. D
5. B
6. A
7. C
8. D
9. A
10. D
11. B
12. D
13. C
14. C
15. D

Chapter 16

Sociology and the Environment

Chapter Introduction

Though I will admit that I should probably be more environmentally aware, I do try to use environment-friendly products and can honestly say that I've never littered in my life. I'm usually too busy to stop and appreciate the beauty that surrounds me, but last October I was driving home from a conference and I got lost. I ended up driving around for hours before I finally stopped and asked for directions. I would have stopped much sooner, but I was absolutely captured by the beauty of the leaves turning colour, and for the first time in far too long, I was really glad that I live in Canada. I suspect that it's much the same feeling that those who live close to the rain forests experience. I can't imagine the destruction of such beauty, yet it is occurring at an uncomfortably rapid pace. Try to take some time from your busy schedule to appreciate the wonders of nature that exist right outside your front door, and imagine how you would feel if, one day, your favourite tree was gone and replaced by a communication tower. It's a very scary thing to think that our natural resources are being depleted, especially when you realize that it takes years to get them back and in some cases, we never will. How often do you think about the environment that you live in? Do you recycle? Do your parents? What about the celebrities who promote environmental causes? Steven Page from the Barenaked Ladies is definitely environmentally friendly, as are the rest of the members of the band. This year he bought a 2004 Toyota Prius hybrid, and I'm wondering if sales of that car will increase because of his purchase. Does celebrity status and the causes they promote influence your behaviour in any way?

In this chapter you'll examine how sociology has dealt with global environmental awareness, concern, and actions. The author briefly outlines the traditional lack of concern with the environment in sociological theory and research and then examines the value conflict in contemporary societies. The value conflict exists between those who favour unlimited economic expansion and technological solutions to human problems and those who embrace a new "ecological" view of the world, in which nature is accorded a central place. You will also learn about four principal areas of sociological inquiry that relate to the environment. When you consider what an environmentally aware society we have become, you'll appreciate how much useful and thought-provoking information this chapter contains.

Jump Start Your Brain!

In this chapter you will learn that:

1. A major focus of the sociology of the environment is the conflict between environmentalists and their opponents in industry and science.

2. Support for environmentalism has remained constant for nearly three decades, with a majority of people generally supportive of environmental values and a young, well-educated, urban, liberal group leading the movement for environmental change.
3. To mobilize the reluctant majority, organizers of the environmental movement develop and spread interpretations of events that play up the possibility of environmental crises.
4. The goals of conserving resources, reducing pollution, and restricting population increase are especially difficult to achieve in the developing world.
5. At the community level, willingness to act on environmental problems rises as trust in authority figures declines.
6. Environmental problems are often contested on the basis of acceptable risk; the definition of what is acceptable risk is strongly influenced by the distribution of power in society, with more powerful individuals and groups better able to determine what is and what is not risky.

Quiz Questions

True or False?

1. A major focus of the sociology of the environment is the conflict between environmentalists and their opponents in industry and science. True or False
2. The goals of conserving resources, reducing pollution, and restricting population increase are becoming easier to achieve in the developing world. True or False
3. In the dominant paradigm of our society, progress is interpreted as the increasing encroachment of civilization on jungles, deserts, frozen tundra, and other "wild" geographic environments. True or False
4. Today, "environmental sociology" has become a catchall for the study of all social aspects of the environment. True or False
5. Schnaiberg (1980) has described a pervasive conflict in advanced industrial societies between the economy and the rising public demand for protecting the environment. True or False
6. At the level of the local community, willingness to act on environmental problems rises as trust in expert institutions increases. True or False
7. Deep ecology is an environmental ethic emphasizing that all species in nature are of equal value. Our experience of nature, deep ecologists claim, should be the foundation for an energetic environmentalism that opposes the present domination by rational science. True or False
8. Beginning in 2001, integrated resource management policy was increasingly expressed in terms that tapped into the discourse of "sustainable development" or "ecological modernization." True or False

9. A major attempt to bridge the differences between the dominant and alternative environmental outlooks can be found in the idea of sustainable development. True or False

10. The "economic-contingency" hypothesis predicts that environmental concern will eventually diffuse throughout all groups. True or False

Multiple Choice

1. The environmental movement has grown tremendously since the _____________. Some members of the middle class in particular are now more personally involved in the environmental movement and raise environmentally conscious children.

   a. 1980s

   b. 1990s

   c. 1960s

   d. 1970s

2. Recent research has suggested that environmental issues rise and fall in the public eye in response to a number of different factors. Which of the following is *not* one of them?

   a. the public cost of increasing environmental concern

   b. the clarity and viability of scientific evidence

   c. the ability of environmental claims-makers to sustain a sense of dramatic crisis

   d. the rise of competing new environmental problems

3. The _____________ comprises a set of beliefs that challenge the centrality of economic growth, technological progress, and the human domination of nature as pillars of society. This paradigm stresses the need to adopt small-scale, decentralized economic and political structures that are in harmony with nature.

   a. dominant paradigm

   b. human-exceptionalism paradigm

   c. alternative environmental paradigm

   d. utopian equality paradigm

4. Support for environmentalism has remained remarkably constant for nearly ___________ years.

   a. 25

   b. 20

   c. 15

   d. 10

5. By the early _____________, stimulated by increased societal attention to urban decay, pollution, overpopulation, resource shortages, and so on, a number of sociologists began at last to study environmental issues.
   a. 1950s
   b. 1960s
   c. 1970s
   d. 1980s
6. Frames are interpretations of events and their meanings. Successful framing involves three kinds of elements. Which of the following is *not* one of them?
   a. analytic elements
   b. diagnostic elements
   c. prognostic elements
   d. motivational elements
7. More recently, environmentalists have been identified as members of a new ____________ drawn primarily from social and cultural specialists.
   a. lower-middle class
   b. middle class
   c. middle-upper class
   d. upper class
8. After studying recycling behaviour across Alberta, Derksen and Gartrell (1993) concluded that the key factor accounting for participation in recycling programs was _____________.
   a. the increase in respondents' environmental awareness
   b. the feeling that respondents' get from knowing that they're helping save the environment
   c. the ease with which respondents can dispose of their waste
   d. the easy availability of curbside pickups
9. In ______________, the conservation movement developed in a different fashion. Environmental initiatives, such as the establishment of national parks and the protection of wildlife, were more likely to be developed by small groups of dedicated civil servants who were able to convince the federal government to take action.
   a. England
   b. the United States
   c. Canada
   d. France

10. Ecofeminism was first coined in _____________ by the French writer Françoise d'Eaubonne, who believes that the oppression and exploitation of women and the domination of the natural environment are part of the same phenomenon.

    a. 1974

    b. 1944

    c. 1984

    d. 1962

11. Deep ecologists believe in a(n) _____________ approach which emphasizes that humans are one species among many on Earth and have no special rights or privileges.

    a. sustainable development

    b. environmental management

    c. anthrobiocentric

    d. biocentric

12. Environmental problems are often contested on the basis of acceptable risk; the definition of acceptable risk is strongly influenced by the distribution of _____________ in society.

    a. prestige

    b. wealth

    c. power

    d. income

13. One of the first and most important efforts to develop a research tool with which to measure an environmental view of the world was _____________ new environmental paradigm (NEP) scale.

    a. Dunlap and Catton's (1979)

    b. Jones and Dunlap's (1992)

    c. Grossman and Potter's (1977)

    d. Dunlap and Van Liere's (1978)

14. History, as it has been taught in our schools, is an account of how the explorers, missionaries, traders, and industrialists rolled back the frontier, "tamed" nature, and brought prosperity to "virgin" lands. The great achievements of the last two centuries, except _____________, all represent a triumph by science and industry over natural hazards and barriers.

    a. the Great Wall of China

    b. the opening of the Panama canal

    c. the completion of the Canadian transcontinental railway

d. the landing of the astronauts on the moon

15. In the 1980s, the threat of global collapse shifted to that of "biosphere crisis," generated by ____________ climatic changes resulting from increased emissions of "greenhouse" gases into the atmosphere.

    a. national

    b. provincial

    c. global

    d. regional

Critical Thinking

1. Greenbaum (1995) has characterized the social bases of environmental concern as "complex and subtle." That's because environmental concern spans a wide variety of subject matters, from species extinction and the thinning of the ozone layer to the contamination of local drinking waters by toxic chemicals. Although it may be possible to isolate general clusters of environmental concern, people may not be very consistent across various issues. Part of the reason for inconsistency is that individual environmental problems affect us in very different ways. With that in mind, how do you think that environmental support could/should be measured? Should there bc varying levels of support? For example, what if you don't recycle, yet you support the preservation of the rain forests? Could your opinion be captured accurately by using a scale that is so rigid?

2. The authors of the best-selling book *The Limits to Growth* (Meadows et al., 1972) forecast that the earth's carrying capacity—that is, the optimum population size that the planet can support under present environmental conditions—would eventually be exceeded. They predicted that, within a century, we would face a major crisis brought on by uncontrolled population growth and rising levels of pollution. Global warming is becoming increasingly evident as we experience warmer and warmer winters—not that I'm complaining. What are your thoughts on this? Do you think that you would think or act differently when/if you have children? To what extent do you care about future generations? Do you think that as an individual, your actions make a difference?

3. Pamela Anderson is a staunch supporter of PETA, and River Phoenix bought acres of land in the rain forests of Costa Rica before his tragic death in 1993. For years celebrities have endorsed products, social movements, and causes of all kinds—with varying success—even though in some cases, a celebrity endorsement does not necessarily improve the sales of the product. In the introduction, I mentioned Steven Page's new car. How influenced are you by the involvement of a celebrity in a worthwhile cause? How will you feel if twice-nominated U2's Bono eventually wins a Nobel Peace Prize?

4. Schnaiberg (1980) argues that consumers are persuaded from early childhood to become part of a dominant materialistic culture in which personal identity depends on material possessions. What, if anything, is wrong with this? Do you think that he is referring to today's society or perhaps societies of the future? It's

time to get honest here—how environmentally friendly are you? Are you faithful in your recycling and composting? When I was younger, we didn't much worry about the environment, but back then we didn't have a lot of pollutants either. Were you raised in an environmentally friendly family? Will you continue the tradition?

Time to Gear Down!

At the conclusion of this chapter, you should be able to discuss or write about the following, without having to rely on the textbook:

1. Sociological interest in the natural environment is quite recent, having first developed in the early 1970s. Sociology's reluctance to embrace the study of the environment reflects its heritage, wherein biology and nature were banished from the discipline in favour of socially based theories of behaviour.
2. A central focus for much of the sociological examination of the environment has been the deep-seated value cleavage between environmentalists and their opponents in industry and science. The latter support a dominant social paradigm that stresses materialism, economic growth, and the human right to dominate nature. In contrast, environmentalists propose an alternative environmental paradigm that emphasizes the need to adopt small-scale, decentralized economic and political structures that are in harmony with nature. This value-oriented environmentalism has found its fullest expression in a number of "ecophilosophies"—deep ecology and ecofeminism—that have recently flourished on the margins of the environmental movement.
3. Support for environmentalism has remained remarkably constant for nearly 20 years. Although the majority of the population is generally supportive of environmental values, a young, well-educated, urban, liberal core has taken the lead in working for environmental change. Most other Canadians will recycle, purchase "green" products, and act positively toward the environment, but only to the extent that such action does not require any real sacrifice in terms of time and money.
4. To mobilize the reluctant majority, environmental-movement organizers develop frames (interpretations of events) that play up the possibility of an impending global collapse as a result of uncontrolled population growth and continued industrial growth. Global warming, expanding holes in the ozone layer, and the worldwide loss of biodiversity are the most recently identified symptoms of the impending crisis. The only solution, it is claimed, is to draw back and ease down, conserving resources, reducing pollution, and restricting population increase. However, these goals are especially difficult to achieve in the expanding economies of the developing world, where the environment is threatened by both unsustainable development and unsustainable impoverishment.
5. At the level of the local community, willingness to act on environmental problems rises as trust in expert institutions declines. This loss of trust is characteristic of neighbourhood-based environmental conflicts, in which citizens typically find the explanations and assurances offered by scientists and other authority figures to be

faulty. Environmental-risk perception and action are also linked to people's participation in local social networks and community affairs.

6. The role of environmental entrepreneurs or claims-makers is vital in moving environmental issues from free-floating concerns to problems that are recognized and acted on by those in power. These promoters, situated in science, environmental organizations, and the media, define such problems as acid rain, global warming, and ozone depletion, package them, and elevate them to action agendas.
7. The social construction of environmental problems does not occur in a vacuum but is shaped by political and economic factors, to the extent that the powerful in society have the ability to act as gatekeepers, determining what is and what is not relevant with respect to the environment. Environmental problems, then, are actively contested, often on the basis of acceptable or unacceptable risk. Social constructionism in the context of power inequality represents a promising sociological route to understanding the environment-society relationship.

Web Links

TreeHugger

http://www.treehugger.com

OK—This is without a doubt one of the coolest websites I have ever been on! It contains awesome videos and all kinds of information relating to virtually everything environmental from a totally different and happening approach. Literally, you could spend hours here.

Canadian Environmental and Sustainable Development Research Capacity—Sociology

http://www.ec.gc.ca/erad/eng/sociology_e.cfm

This is the home page of Environment Canada, but this link will take you to a directory of information from a host of Canadian sociologists that will update you on their research interests in the area of sustainable development in Canada. And yes, there's even an "Environmental Trivia" link!

Global Recycling Network

http://grn.com

Global Recycling Network is an electronic information exchange that specializes in the trade of recyclables and the marketing of eco-friendly products. This site will tell you literally everything you want to know about recycling goods—from paper products to hockey equipment!

People for the Ethical Treatment of Animals

http://www.peta.org

This is the home page of PETA, and it is a happening site. Check out all the celebrity members!

Build Your Own Solar Car

http://www.re-energy.ca/t_solarelectricity.shtml

Nope, I'm not kidding. Not only does this website explain how solar energy works, it provides a link for you to actually send away for the materials to build your own solar car.

Critical "Linking" Question

Vroooommm! Check out the Build Your Own Solar Car website, and see if you could follow the directions to actually build your car. The one that's listed here looks pretty basic and is relatively cheap—I can't imagine driving it to school every day, but this site also provides information on Solar Car Challenges and races too!

Solutions

True or False?

1. T
2. F
3. T
4. T
5. F
6. F
7. T
8. T
9. T
10. F

Multiple Choice

1. C
2. A
3. C
4. B
5. C
6. A
7. B
8. D
9. C
10. A
11. D
12. C
13. D
14. A
15. C

Chapter 17
Health and Aging

Chapter Introduction

I love old people—probably because I never knew my grandparents and always feel ripped off that they died before I grew up. Now, whenever I get to meet anyone's grandparents, it's quite embarrassing: I make a big fuss. When I was doing my undergraduate degree I used to volunteer at a nursing home and I would sit and listen to the residents' stories for hours. Once I took the old guys fishing and snuck a 12-pack in the lunch cooler. We spent the whole day just fishing, laughing, and telling stories. It was one of the best days of my life. I like to think it made a difference in their lives too—though I'm sure they're all in heaven now. I think one of our country's greatest resources lies in our senior population, and I was thrilled about the changes to the mandatory retirement laws. In my life, I've seen so many older people forced—yes, I hate that word too—into retirement who were just not ready to give up that active and productive part of their life. For some, it led to depression and much loss of self-esteem. But I also know many seniors who could hardly wait for retirement, who planned for it and enjoy it immensely. As a senior, whether you decide to work, relax, or travel in your "golden years" will very much depend on your health as detailed here. As young students, I'm wondering if you give much thought to your older years and if you're taking good care of your mind and body so you get to enjoy your life to the fullest.

In this chapter you'll learn about health as people begin to get older, and then the authors make some connections between age and health and social inequalities. Lastly, they take a good look at health-related issues within Canada and how they fare compared with international and global health trends.

Jump Start Your Brain!

In this chapter you will learn that:

1. People are living longer and healthier lives currently than in the past. Even though this is a triumph over the alternative, the increasing size of the older population tends to be considered a social problem, one that has negative implications for demands on the nation's resources.

2. Health is about more than the presence or absence of disease and, in contrast to generally accepted beliefs, not all aspects of health decline as we age.

3. In Canada, significant inequalities in health are evident in association with social-structural location, as indicated by such factors as socioeconomic position, gender, ethnicity, race, and age.

4. Among older adults, the most common forms of health care consist not of medical and hospital care but rather of self-care and informal care, provided primarily by family members. Of those providing care, most are women.

5. Recent years have seen increased emphasis placed on privatization and profit-making within Canada's health-care system, particularly when it comes to services that are most important to older adults. The implications are that those most in need of care will be the least likely to access it and the most likely to have to rely on themselves and their family members for care.

Time to Gear Down!

At the conclusion of this chapter, you should be able to discuss or write about the following, without having to rely on the textbook:

1. The health and longevity of the Canadian population has been increasing in recent decades. Although this could be considered a positive feature of modern society, instead, it is frequently regarded as a social problem. Older adults are seen as overly dependent on health and social services and, consequently, responsible for the crisis in our health-care system and depriving other generations of their fair share of the nation's limited resources. However, the older population is not the primary source of increases in health-care costs. Increases in the numbers in the older population who may need to access social programs appear to be offset by declines in the proportion of younger people in the population. Sociologists have referred to this type of thinking as apocalyptic demography and suggest that we consider whose interests are served by promoting this view of an aging society.
2. Health is about more than the presence or absence of disease and, in contrast to generally accepted beliefs, not all aspects of health status decline as we age. Older adults, as a whole, assess their health in positive terms and have good mental health and social well-being.
3. Despite the overall picture of good health in old age, in Canada, significant inequalities in health also tend to be evident in association with social structural location, as indicated by such factors as socioeconomic position, gender, race, and ethnicity. Past increases in health and longevity have been concentrated in more advantaged social groups. Aboriginal and immigrant seniors, older women, and the poor continue to experience major health concerns. It is generally believed that tomorrow's older adult population will be better educated and in better health than the older adults of today. Evidence that, currently, poverty in Canada is increasing, inequalities are widening, and the situation may not improve in the near future.
4. Health-related inequalities have been attributed to individual health behaviours and lifestyle factors, psychosocial factors such as stress, and material conditions and resources. Research findings suggest that economic circumstances, living conditions, and other material factors are the most important and that, to a large extent, health-related lifestyles (e.g., diet, exercise, drinking patterns) and stress levels tend to reflect these conditions.
5. We tend to equate health care with medical care generally provided by physicians located in clinics or hospital settings. However, among older adults as well as others, the most common forms of health care consist not of medical and hospital care, but rather of self-care and of informal care, generally provided by family members. When it comes to formal health-care services, home care is particularly important to the

health and well-being of older adults, yet it is not included within Canada's nationally insured health-care system (medicare).

6. Increased emphasis is being placed on privatization and profit-making within Canada's health-care system, particularly when it comes to services that are most important to older adults. The implications are that those most in need of care will be least likely to access it and the most likely to have to rely on themselves and family members for care.

Quiz Questions

True or False?

1. Physicians are the gatekeepers to our health care system and to public health-care dollars. They control access to hospitals, to medical tests, and to prescription drugs. True or False
2. To blame the victim is the tendency to hold victims of various negative events or situations (e.g. crime) completely or partially responsible for what has happened to them, when in reality, the causes are embedded in societal structures. True or False
3. In Canada, because we have socialized medicine, health care is a provincial government responsibility. True or False
4. People with a university degree often feel healthy and function well late into their 60s, 70s, and 80s, whereas those with less education do not. True or False
5. Research findings suggest that not all improvements in health and longevity experienced over the past decades have been equally shared: inclines in death rates in the past few decades have been more rapid in some segments of the population, such as higher-income groups, than in others. True or False
6. The ethnocultural composition of our senior population is heavily influenced by the immigration policies that were in effect during each period of immigration. Consequently, There are more foreign-born individuals within Canada's younger population than there are within the older population. True or False
7. Women are often the "kinkeepers" in society, whereas men tend to rely on their wives for social connectedness. True or False
8. Deaths in old age, and remember most deaths in Canada now occur in old age, usually result from chronic degenerative diseases. True or False
9. With 13 percent of its population currently aged 65 and over, Canada is not considered to have an old population. True or False
10. Not until their 70s do men and women have the approximately same number of years left to live. True or False

Multiple Choice

1. Today, most Canadians can expect to live to an old age, barring accidents and wars. People can now expect to live more than _____________ years longer than if they had been born in 1921.

a. 5

b. 10

c. 15

d. 20

2. _____________ (a research area that studies the characteristics of populations and the dynamics of population change) considers a country to be old if at least 10 percent of its population is age 65 and older.

   a. Gerentology

   b. Ageism

   c. Demography

   d. Population control

3. The reasons that women outlive men are not known. It is believed that there is a(n) ___________ component, but current gender differences are determined at least in part by social and economic forces.

   a. biological/genetic

   b. emotional/psychological

   c. marital/parental

   d. religious/racial

4. When many of today's senior's were immigrating to Canada, it was primarily English- and French-speaking individuals from Europe and other Western countries who were granted permission to enter. Indeed, in 1996, only _____________ percent of seniors were members of visible minority groups.

   a. 2

   b. 4

   c. 6

   d. 8

5. The most prevalent form of dementia is Alzheimer's disease. Canada's only national study to date to establish the prevalence of dementia revealed that between _____________ and _____________ of seniors have this disease.

   a. 3%; 5%

   b. 6%; 8%

   c. 9%; 11%

   d. 12%; 14%

6. Although education is important to health, it has been estimated that approximately 23 percent of premature mortality (i.e. potential years of life lost) among Canadians can be linked to _____________ differences.

a. race-related
b. income-related
c. divorce-related
d. gender-related

7. Three types of factors have been linked to structured inequalities on the one hand and poorer individual physical and mental health on the other. Which of the following is *not* one of them?

a. individual health behaviours and lifestyles
b. ratings on the calculated risk factor index
c. material conditions and resources
d. social psychological resources

8. It has been estimated that in virtually all developed countries, irrespective of what type of health-care system they have and whether or not they have universal health care, approximately _____________ of all care that is provided to seniors comes from the informal network, primarily from family members.

a. 40%
b. 50%
c. 65%
d. 75%

9. _______________ is the major predictor of population aging until a population reaches a life expectancy at birth of 70 years of age, at which point almost all young persons survive.

a. Fertility
b. Disease
c. Sanitation
d. Medical technology

10. Within the seniors population, ethno-minorities constitute a relatively small but important group. Approximately __________ seniors immigrate to this country every year, most often as family-class immigrants, and are usually of non-European origin.

a. 3000
b. 6000
c. 9000
d. 12,000

11. The extent to which health problems interfere with day-to-day functioning or limit activities is typically referred to within gerontology as _____________.

a. chronic conditions

b. psychosomatic disorder

c. pain

d. functional disability

12. According to proponents of ______________, stress itself can not only cause poorer health, but it can also lead people to smoke, consume too much alcohol, eat too much or too little, or sleep too little, all of which can also have negative impact on health.

a. hierarchy stress perspective

b. social causation hypothesis

c. social selection hypothesis

d. blaming-the-victim theory

13. According to Canada's Royal Commission on Aboriginal Peoples (1996), the life expectancy (at birth) of registered Indians in Canada is ____________ years shorter than that of non-Aboriginal Canadians.

a. two to four

b. four to six

c. seven to eight

d. eight to nine

14. According to the ____________, Western nations are successfully postponing the age of onset of chronic disability so that more and more people experience it later and later in life.

a. delaying disability hypothesis

b. compression of mortality hypothesis

c. compression of morbidity hypothesis

d. postponement of mortality hypothesis

15. A minority of seniors, particularly the frail, poor, and old-old, are vulnerable to social isolation. However, the majority are embedded within ____________, characterized by mutual and close intergenerational ties, responsible filial behaviour and contact between generations.

a. modified kinship family networks

b. well-intended family networks

c. modified extended family networks

d. modified nuclear family networks

Critical Thinking

1. In the beginning of this chapter, one of the authors writes about a headline she reads: "Can We Prepare for Life at 100?" (Ibbitson, 2005). OK—how insane is that? I can barely begin to prepare for the day each morning. Seriously though—you're likely

very young and not yet even thinking about life at 50 let alone 100. Sit for a minute and consider what it will be like to live for another 80 years. Have you started an RRSP account? Are you taking good care of your body *and* your mind?

2. Please check out Box 17.2 on page 453 before you attempt to answer this question and take a pretty good look at it too. I had to read both lists a couple of times before I really got a good sense of what was going on. Is it my imagination or are these two lists completely different? Which list makes more sense to you? Which one would you have an easier time adhering to? Try to make a list that combines aspects from each list so that you've got sort of an "uber-list" that might work for everyone.
3. I found it quite interesting to read about how the concept of old age came to be and even a bit shocking to learn that we still use that definition today. Read Box 17.1 on page 444 and see if you agree with present-day standards. Your mission for this question however, is to try and figure out how you would define young age! What measures would you use? How would you ensure that it stood the test of time?
4. The relationship between health and aging is a matter of both common sense and paradox. Is good health a problem or is bad health a problem? With the rising costs of organic foods and good nutrition, do you think it costs more to stay healthy or to fix yourself when you're sick? What's your strategy to good health? Or do you leave it to fate and believe that regardless of how well you take care of yourself, when you're number's up there's not much you can do?

Web Links

The Death Clock

http://www.deathclock.com

If you believe in this sort of thing. Fiddle with the mood button and see how it extends your life. Then fiddle with the smoking and non-smoking button—see anything interesting?

Find Your Fate

http://www.findyourfate.com/deathmeter/deathmtr.html

OK—I couldn't resist—same sort of thing, but look how at how much more detail this site goes into. C'mon—you know you want to—try both and compare dates!

Body Mass Index

http://www.hc-sc.gc.ca/fn-an/nutrition/weights-poids/guide-ld-adult/bmi_chart_java-graph_imc_java_e.html

Use this site to calculate your BMI—the scientific standard that's used to determine acceptable weight levels—and don't be surprised at the results! Navigate the rest of the site for excellent nutrition and health tips!

International Development Research Centre

http://www.idrc.ca/en/ev-1-201-1-DO_TOPIC.html

Phenomenal find! Definitely watch the video clip—this global program is all about building better health systems all over the world, to create a better, healthy population—it's totally amazing what our most valuable resource—people—can do!

Global Aging

http://www.globalaging.org/index.htm

A really neat way for seniors—and non-seniors—to keep tabs on what's going on in the world. This American-based, not-for-profit website offers a global look at age-related issues in the world.

Critical "Linking" Question

Check out the International Development Research Centre website and take a really good look at some of the projects. How successful do you think they will be? How much of an impact will they make on the overall health of the population? When you watched the opening video, how did it make you feel? Check out the winners of the 2006 photo contest at this link: http://www.idrc.ca/en/ev-99119-201-1-DO_TOPIC.html.

Solutions

True or False?

1. T
2. F
3. F
4. T
5. F
6. F
7. T
8. T
9. F
10. F

Multiple Choice

1. D
2. C
3. A
4. C
5. B
6. B
7. B
8. D
9. A
10. B
11. D
12. A
13. C
14. C
15. C

Chapter 18
Politics and Social Movements

Chapter Introduction

I almost cringe when I admit that I am not a politically minded person. I know that I should be and I know how important it is to be politically aware of the "goings on" in our country, but the truth of the matter is that I'm just not as keen in this area as I could or more importantly should be. On the other hand, I also don't moan and complain about the government a lot, because I feel as though I don't have a right to do so, since I don't fully participate in the politics of it. Have you ever participated in a rally or peace protest? Are you a registered voter and have you exercised your right to vote? Can you see yourself leading a demonstration outside the Parliament buildings? Politics and social movements often go hand in hand and both are integral parts of studying sociology. If you've ever been involved in or volunteered for a political campaign, you probably have a good idea of the massive amount of time, work, and money that goes into every election. Are you a politically minded person who follows politics closely, or are you someone who is relatively apathetic? It seems to me that people fall into one of two groups when it comes to politics. On one side there are those who are avidly political and aware of what's going on in Canada and around the world, and then there are those that fall into the less enthusiastic or apathetic category. These are the people who feel that they don't know enough about politics to make an informed choice on a candidate and because they don't want to spoil a ballot, they don't vote. Of course, I could be totally wrong on this one, but this is the way that it appears to me. How would you categorize Canadians in terms of political awareness? Does it surprise you that the United States has a very low turnout for voters? I've always considered the States to be an extremely patriotic nation, but results from recent research don't indicate that.

In this chapter, you'll learn that the power of a group may be widely recognized as legitimate or valid under some circumstances. If it is, raw power becomes legitimate authority. Weber's typology of authority has three bases: traditional, legal rational, and charismatic. The people who occupy the command posts of institutions are generally seen as authorities. Under other circumstances, however, power flows to non-authorities. This undermines the legitimacy of authority. In this case, non-authorities form social movements or collective attempts to change part or all of the social order. They may riot, petition, strike, demonstrate, and establish pressure groups, unions, and political parties in order to achieve their aims. Additionally, you'll learn how politics has developed over the past 300 years and what you can reasonably expect in the near future.

Jump Start Your Brain!

In this chapter you will learn that:

1. The level of democracy in a society depends on the capacity of citizens to influence the state through their support of political parties, social movements, and other groups. That capacity increases as power becomes more widely distributed in society.

2. The degree to which power is widely distributed influences the success of particular kinds of parties and policies.
3. People sometimes riot, strike, and take other forms of collective action to correct perceived injustices. When they do so, they are participating in social movements.
4. People are more inclined to rebel against the status quo when they are bound by close social ties to other people who feel similarly wronged and when they have the money and other resources needed to protest.
5. For social movements to grow, members must make the activities, ideas, and goals of the movement congruent with the interests, beliefs, and values of potential new recruits.
6. The history of democracy is a struggle for the acquisition of constantly broadening citizenship rights.

Quiz Questions

True or False?

1. The degree to which power is widely distributed influences the success of particular kinds of parties and policies. True or False
2. For social movements to grow, members must make the activities, ideas, and goals of the movement congruent with the interests, beliefs, and values of potential new recruits. True or False
3. People sometimes riot, strike, and take other forms of collective action to correct experienced injustices. Whey they do so, they are participating in social movements. True or False
4. Civil citizenship is the right to free speech, freedom of religion, and justice before the law. True or False
5. With the exception of Switzerland, Canada has the lowest voter turnout of any rich democracy in the world. True or False
6. Some Marxists, known as "instrumentalists" deny that elites enjoy more or less equal power. True or False
7. Pluralist theory is one interpretation of the relationship between state and civil society. True or False
8. The history of democracy is a struggle for the acquisition of equality. True or False
9. Elite theory maintains that well-to-do people consistently have more political influence than people who are less well-to-do and that society is therefore not as democratic as it is often portrayed. True or False
10. State-centred theory suggests that social movement formation and success depend on how powerful authorities are compared with partisans of change. True or False

Multiple Choice

1. _____________ is the process by which individual interests, beliefs, and values either become congruent and complementary with the activities, goals, and ideology of the movement or fail to do so.

   a. Relative deprivation

   b. Resource mobilization

   c. Social alignment

   d. Frame alignment

2. According to Weber's typology, which of the following is *not* a base of authority?

   a. traditional authority

   b. charismatic authority

   c. divine authority

   d. legal-rational authority

3. The _____________ is a set of institutions that formulate and carry out a country's laws, policies, and binding regulations.

   a. House of Commons

   b. Senate

   c. Legislature

   d. State

4. From studies of political participation in Canada, many surveys show that political involvement decreases with ________________.

   a. income

   b. education

   c. social class

   d. geographic location

5. Only in the United States do individual citizens have to take the initiative to go out and register themselves in voter registration centres. As a result, the United States has a proportionately smaller pool of eligible voters than the other democracies do. Only about _____________ of American citizens are registered to vote.

   a. 23%

   b. 34%

   c. 48%

   d. 65%

6. _____________ is an intolerable gap between the social rewards people feel they deserve and the social rewards they expect to receive.

a. Relative deprivation

b. Frame alignment

c. Resource mobilization

d. Power-balance

7. Social movements often used their power to expand the rights of citizens. Which is *not* a stage in this process?

   a. civil citizenship

   b. legal citizenship

   c. social citizenship

   d. political citizenship

8. There are three elements to the State. Which one is *not* an element?

   a. executive

   b. legislature

   c. law enforcement

   d. judiciary

9. The peak year of strike activity in Canada was ___________. In that year, 17.3 strikes took place for every 100,000 non-agricultural workers in the country.

   a. 1899

   b. 1919

   c. 1929

   d. 1945

10. ____________ enable workers to effectively bargain with employers and governments for improved wages, working conditions, and social policies.

    a. Collective agreements

    b. Benefit packages

    c. Unions

    d. Mediators

11. ____________ theorists teach us that, despite the concentration of power in society, substantial shifts in the distribution of power do occur, and they have discernible effects on voting patterns and public policies.

    a. Power-balance

    b. Elite

    c. State-centred

    d. Pluralist

12. Members of disadvantaged racial minority groups, especially ________________, are less likely to register than whites.

    a. African Americans

    b. Asians

    c. Aboriginals

    d. East Indians

13. ________________ is based on the idea that social movements can emerge only when disadvantaged people can marshal the means necessary to challenge authority.

    a. Relative deprivation theory

    b. Resource-mobilization theory

    c. State-centred theory

    d. Power-balance theory

14. In the world's rich countries, a strong ________________ did much to promote the growth of democracy in its early stages.

    a. bourgeoisie

    b. class consciousness

    c. proletariat

    d. social citizenship

15. Research shows that in Canada since World War II, strike activity has been high when three factors were in place. Which of the following is *not* one of the factors?

    a. unemployment low

    b. union membership high

    c. economy stable

    d. governments generous in their provision of social welfare benefits

Critical Thinking

1. In Canada and other democracies, working-class and poor people are less likely to vote than members of the middle and upper classes. They are also less likely to contact their elected representatives, contribute to election campaigns, and run for office. Only people with high incomes, substantial wealth, and postsecondary education have the time, money, and social connections required for the most intense forms of political engagement. What does this tell you about our elected officials? How does this class bias affect how we are governed?

2. By now you've probably learned a whole lot about Canada and many other countries that make our world abound with diversity. In terms of politics and government, how democratic is the Canadian state? Does the interaction between state and civil society ensure that every citizen have a roughly equal say in the determination of law and

policies? Do you think that some citizens are just a little more equal than others? Or do you think that some citizens are a lot more equal than others?

3. Certainly, in any given election, the entire eligible population does not vote. Have you ever voted in a federal, provincial, or municipal election? If not, perhaps you have voted in a school election. Some people think that their vote can't change a thing, and others feel that they don't know enough about politics to cast their vote, so they don't bother voting at all. What measures could be taken to ensure greater voting participation? Do you think that online voting could or should be implemented? How would this happen? What potential pitfalls do you see?

4. If you have not already done so, would you participate in a rally or protest? How strong would your beliefs have to be in order to formally protest against something? What type of cause are you more likely to support? Does the type of support you offer (e.g., time or monetary donation) depend on the cause that you're willing to support?

Time to Gear Down!

At the conclusion of this chapter, you should be able to discuss or write about the following, without having to rely on the textbook:

1. Democracy involves a two-way process of control between the state (the set of institutions that formulate and carry out a country's law, policies, and binding regulations) and civil society (the private sphere, consisting of social movements, political parties, etc.).

2. The level of democracy in a society depends on the capacity of civil society to influence the state through citizen support of social movements, political parties, and other groups. That capacity increases as power becomes more widely distributed in society.

3. Although pluralists are correct in noting that democratic politics is about negotiation and compromise, they fail to appreciate how advantaged groups tend to have more political influence than others.

4. Although elite theorists are correct in noting that power is concentrated in the hands of advantaged groups, they fail to appreciate how variations in the distribution of power influence political behaviour and public policy.

5. While power-balance theorists focus on the effect of changes in the distribution of power in society, they fail to appreciate what state-centred theorists emphasize—that state institutions and laws also affect political behaviour and public policy.

6. The degree to which power is widely distributed influences the success of particular kinds of parties and policies. Widely distributed power is associated with the success of labour parties and policies that redistribute wealth.

7. Research does not support the view that social movements emerge when relative deprivation spreads.

8. Research does suggest that people are more inclined to rebel against the status quo when they are bound by close social ties to many other people who feel similarly wronged and when they have the money and other resources needed to protest.
9. For social movements to grow, members must engage in frame alignment, making the activities, goals, and ideology of the movement congruent with the interests, beliefs, and values of potential new recruits.
10. The history of democracy is a struggle for the acquisition of constantly broadening citizenship rights—first the right to free speech, freedom of religion, and justice before the law, then the right to vote and run for office, then the right to a certain level of economic security and full participation in the life of society, and finally the right of marginal groups to full citizenship and the right of humanity as a whole to peace and security.
11. In the developing world, social movements have focused less on broadening the bases of democracy than on ensuring more elemental human rights, notably freedom from colonial rule and freedom to create the conditions for independent economic growth. In some cases these movements have taken extreme, antidemocratic forms.

Web Links

Elections Canada

http://www.elections.ca/intro.asp?section=gen&document=index&lang=e

Once again, I've found a very groovy site. This find contains scads of information about elections—past and present—and voting patterns in Canada. You would never in a million years think that it would be cool to check this out, but trust me, it is!!!

Political History and Evolution: Your Canadian Election Headquarters

http://www.elect.ca/evolution.php

A great site on the political history of Canada, and the whole site is great to explore! Test your knowledge on the Canadian Trivia Quiz.

Social Movements

http://www.wsu.edu/%7Eamerstu/smc/smcframe.html

This is just a very cool site with links that can connect you to a veritable cornucopia of social movements. Truly, they have links to social movements that I would have never thought about or studied. Definitely worth a cruise.

The Women's Movement

http://www.canadianencyclopedia.ca/index.cfm?PgNm=TCE&Params=A1SEC830180

Hop on this link and learn more about hows and whys of the women's movement in Canada. You'll be surprised at what you've learned and what women have managed to accomplish!

Democracy Watch Homepage

http://www.dwatch.ca

Democracy Watch is Canada's leading citizen group advocating democratic reform, government accountability, and corporate responsibility. Check out some of the articles under the "What's New" link, and see if you think their reporting is fair and unbiased.

Critical "Linking" Question

Take a "surf" on the Elections Canada website and check out all the really cool facts. What do you think accounts for all the differences in voter turnout over the years? Do you think that the games feature are really going to attract younger voters? What's your take on Aboriginal voters—are their voter patterns different than those of non-Aboriginal people?

Solutions

True or False?

1. T
2. T
3. F
4. T
5. F
6. T
7. T
8. F
9. T
10. F

Multiple Choice

1. D
2. C
3. D
4. C
5. D
6. A
7. B
8. C
9. B
10. C
11. A
12. A
13. B
14. A
15. C

Chapter 19
Globalization

Chapter Introduction

I can't remember the first time that I actually thought of the world as a global community—can you? I can tell you quite honestly that I was more than a little creeped out after reading the introduction of this chapter and learning that Ronald McDonald is the second-most-recognized figure in the world. For me, that just seems wrong on so many levels, it's not even funny. How often do you spend time thinking about the relationships that exist all over the world? It amazes me that technology has brought us to a point where people can simultaneously design and manufacture a piece of machinery when they are all in different countries. I can clearly remember my first conference call and I know it was a very different experience for me. I guess I'm a visual person, and I had trouble following the conversation without being able to see with whom I was speaking. Do you drive a car that was built in another country? I once taught a student who refused to wear any clothing that wasn't made in Canada. I thought very highly of his dedication and commitment to the Canadian textile workers. This was well before the Kathie Lee Gifford scandal and the press coverage that surrounded that issue. We spoke about it at some length, and he admitted that his selection in clothing was limited and he was often tempted to become a "victim of fashion" but he never did. His loyalty impressed me greatly, especially since the clothes that he had to buy were all union-made and often were much more expensive than other lines of similar clothing. I think that, for him, what was on the inside of his body was much more important to him than what was on the outside. In any event, that student reminds me that occurrences far away from us can still affect us close to home—I'm hoping that you'll remember C. Wright Mills here and the sociological imagination. It's amazing how something that happens a million miles away can affect the way that we live our everyday lives and, if you stop to think about it, it's amazing how often it happens.

In this chapter you'll realize a more sophisticated sense of what globalization really means, and hopefully understand how different political and economic interests struggle to promote their own brand of globalization. By seeing the world as a "global village" you'll see that the future of the world depends on the shape and direction of global agendas.

Jump Start Your Brain!

In this chapter you will learn that:

1. *Globalization* is a term used to describe how the world "shrinks" as capital, ideas, corporations, commodities, and workers rapidly cross vast distances. You will learn about the complexities of globalization, a term much used and abused, and acquire a working definition of it.

2. Globalization is not a distant force but a real-world phenomenon that affects you daily as you eat, drink, work, and go shopping. By the end of the chapter, you will better understand how globalization relates to you as a consumer, a worker, and a citizen.
3. Globalization is dynamic, because it is contested by different social forces promoting different kinds of globalization. This chapter will give you a conceptual framework that will allow you to appreciate the "top-down" forces of globalization that promote global capitalism and the "bottom-up" forces demanding greater democracy, social justice, and sustainability in the global system.

Quiz Questions

True or False?

1. International consumerism is a social, economic, and political process that makes it easier for people, goods, ideas, and capital to travel around the world and an unprecedented pace. True or False
2. Although the "Three Sisters" influence state policies throughout the world, not all states are equally affected by these institutions. Neoliberal thinkers in Washington acknowledge that the United States acts like an empire, although they argue that it uses its power benevolently to promote peace and democracy throughout the world. True or False
3. Half the world's people still make a living off the land and subsist mainly on what they produce themselves. True or False
4. Although it was unimaginable to promote a "sweatshop-free" clothing line in the early 1990s, the popular label Roots has proven that it is possible to run a successful business without the use of sweatshop labour. True or False
5. Trade experts and environmental groups warn that agriculture is largest contributor of greenhouse gas when food production and distribution chains are taken into account, and they suggest that the global food chain system represents the biggest environmental challenge facing humanity. True or False
6. The world political system has developed as the empire of a super-state as well as a single world state. It has thereby become a forum for global consciousness, where opinions are formed and conflicts resolved by hammering out common global interests. True or False
7. Immanuel Wallerstein (1974) has made the development of the world economy the theme of his world-system theory, which considers economic relations only from the perspective of the world as a whole, with states representing merely one force among many. True or False
8. The global spread of consumerism has been criticized as a form of cultural imperialism and is often associated with liberal values around sexuality, feminism, and secularism. True or False

9. The rise of financial capital has been labelled "casino capitalism" since financial speculators, like casino gamblers, stand to make or lose millions of dollars in short periods. True or False

10. Not all people and ideas have access to channels of globalization like the Internet or even the telephone. Inequality of access to means of communication is known as the technological division of labour. True or False

Multiple Choice

1. One term for the shrinking-world phenomenon is _____________, which suggests that we are no longer slowed down by long distances.

   a. the global village

   b. space-time compression

   c. new age technology

   d. globalization

2. Due to pressure to meet the demands of very powerful international financial institutions, some critics argue that states have become less oriented to meet the demands of citizens. The result in a _____________ in which ordinary citizens are disenfranchised from the process of governance.

   a. loss of trust in the governing bodies

   b. decreased voter turn out

   c. democratic deficit

   d. state of false consciousness

3. The fair-trade movement is one of the leading proponents of subverting corporate logos to disrupt global commodity chains, arguing that the producers should be paid a fair price rather than the free market price. The fair-trade movement has paid special attention to _____________.

   a. coffee

   b. clothing

   c. beef

   d. techno-gadgets

4. Globalization has placed _____________ in a difficult position: wages are too low to alleviate poverty rates yet too high to continue to attract low-cost manufacturing.

   a. Mexico

   b. Vietnam

   c. China

   d. Thailand

5. _____________ of the world's population (1.2 billion people) living in the industrialized developed world currently consume _____________ of the world's resources and create 75 percent of all waste and pollution.

   a. 5%; one-quarter

   b. 10%; one-eighth

   c. 20%; one-third

   d. 30%; one-half

6. The average North American meal travels more than _____________ kilometres to reach your dinner table, so there's a good chance that your burger and fries have been globalized.

   a. 1200

   b. 1600

   c. 3200

   d. 3800

7. Voices from less developed countries often speak of a(n) _____________ in which skilled professionals leave their homelands to seek better opportunities in developed countries, a trend thought to cost India US$2 billion a year.

   a. Einstein decline

   b. global brain drain

   c. neurological deficit

   d. global brain strain

8. _____________ are against the neoliberal forms of globalization that put capital mobility and profits before people's basic needs and they criticize the powerful economic, political, and military influence of transnational corporations and the United States government. .

   a. Casino capitalists

   b. Cultural imperialists

   c. Bottom-up globalizers

   d. Top-down globalizers

9. Governments have reacted in different ways to the global competition to create jobs and attract corporate investment. In the less developed countries, states have set up _____________ where special financial deals are used to lure corporations to set up shop and provide jobs.

   a. legalized tax shelters

   b. export processing zones

   c. specialized work permits

d. job specific labour pools

10. Today, an estimated 60 percent of McDonald's corporate profits come from its international division. McDonald's opens a new restaurant every ____________ and is the world's largest user of beef.

a. 17 hours

b. 24 hours

c. 2 days

d. week

11. Although the global economy has made a portion of the world's population wealthy, a large population of the world's people (at least ____________) are considered poor (and live on less than US$2 a day).

a. 15%

b. 25%

c. 35%

d. 50%

12. ____________—understood as a way of life in which identity and purpose are oriented primarily to the purchase and consumption of material goods—is currently being exported to the world's middle and working classes.

a. Globalization

b. Commodification

c. Capitalism

d. Consumerism

13. Global corporations are producing more things than the world's consumers can afford to purchase. Certain phenomena have changed the way corporations look and operate. Which of the following is *not* one of them?

a. the creation of a global economy

b. the digital divide

c. ruthless competition in the goods and services sector

d. the rise of financial capitalism

14. Top-down globalization has been dominated by ____________ economic policies, which have become prevalent in both rich and poor countries in the past 25 years.

a. conservative

b. democratic

c. radical Marxist

d. neoliberal

15. "Globalization" was coined in the late ____________ and today there are thousands of globalization books, conferences, university courses, and references in newspapers and magazines, many of which contradict each other.

    a. 1960s

    b. 1970s

    c. 1980s

    d. 1990s

Critical Thinking

1. In the opening of this chapter, we learn that in 2001 six vegetarians from British Columbia moved to sue McDonald's after it was revealed that McDonald's french fries use beef fat for "flavouring." After hearing the announcement, vegetarian activists in India held demonstrations and attacked a McDonald's in Bombay, demanding that McDonald's leave the country. McDonald's settled the lawsuit in 2002 by agreeing to donate $10 million dollars to Hindu and other consumer groups. What do you think of this? Compared to the net worth of the McDonald's chain, is $10 million dollars really going to make a difference to their operating costs? What about the activists? Do you think that they "sold out?" If you believed in a cause, would you be inclined to fight to the death, or would you be tempted to settle for cash?
2. Bottom-up globalizers have reacted to the growth of global corporations in various ways. For example, the 1990s saw the emergence of an anti-sweatshop movement in North America after poor working conditions in the garment industry were exposed. Of particular importance was the Kathie Lee Gifford controversy, which revealed that her Wal-Mart clothing line was produced by child labour and involved human rights issues. Major brand names, among them Nike, Starbucks, McDonald's, and Shell Oil, are also principal targets. Do you think that the conditions under which many labourers work would have gone on unnoticed if not for Gifford's celebrity status? To what degree do you think these working conditions still exist in the world? Was Kathie Lee hailed as a scapegoat for many others that engage in the same practice, but are just lucky enough not to get caught?
3. Table 19.2 on page 491 of your text exemplifies globalization and space-time compression, but also sheds light on the stark reality of differences between those areas in the world that have access to technology and those that don't. I very rarely watch television and I don't even own a cell phone, but I do all my banking online and cannot imagine my life without either of my two home computers. Of the four domains outlined, which are the most and least important to you? Which one(s) could you live without? If you could keep only one, which would it be?
4. Right off the bat, this chapter asks a very important question—what do we mean by globalization? I have to tell you, it's a pretty good question indeed. What does globalization mean to you? Is it one of those words that you sort of like to banter about but really don't know the exact meaning of? How would you define it so

that it makes sense and has meaning for you? Does the fact that Ronald McDonald is the second-most-recognized figure in the world influence seem scary to you at all? I will admit that I'm somewhat comforted by the fact that Santa Claus came in first, but seriously, Ronald McDonald? What does this tell us about the importance our society places on our real leaders, or people of great import?

Time to Gear Down!

At the conclusion of this chapter, you should be able to discuss or write about the following, without having to rely on the textbook:

1. Globalization effectively shrinks the world; workers, commodities, ideas, and capital cross distances more quickly. Sociologists use the term *space-time compression* to describe this process.
2. Globalization processes have generated contradictory outcomes that benefit some groups but have also been linked with poverty, economic marginalization, democratic deficits, and the digital divide.
3. Developments in information technology have facilitated the economic integration of financial markets. Consequently, global flows of financial capital are much bigger than global flows of tangible goods and productive capital.
4. Corporations in the global era have become much bigger and are under pressure to become more competitive in the global marketplace. Local or even national competitiveness are no longer seen as sufficient for economic survival.
5. Politically, the globalization era has witnessed the creation of new international institutions of governance, like the IMF, World Bank, and WTO, which have diminished the power and sovereignty of some states.
6. Globalization processes have allowed communities around the world to gain knowledge of the injustice and suffering inflicted by the global economy. Such awareness has inspired efforts to increase social justice in the global system. These efforts are known as "globalization from below."
7. Most of the goods Canadians consume connect them to workers and production processes thousands of kilometres away. The globalization process is almost impossible to escape given the extent of global commodity chains.
8. The period of globalization is associated with a shift in manufacturing employment out of the more developed countries. Competitive pressure is driving corporations to seek the lowest wages possible in the less developed countries (the so-called race to the bottom).
9. Global ecology is not something that exists separately from globalization processes. It is connected to the actions of citizens, consumers, workers, and states.

Web Links

One World

http://www.oneworld.ca

This great website showcases progressive social movements in Canada and way beyond our borders. Just by clicking your mouse you can learn about social justice, human rights, and sustainable development anywhere in the world—one of my best finds yet!

Landscapes of Global Capital

http://it.stlawu.edu/~global/pagesintro/mapfive.html

Wow! This site is mind-boggling—I've never seen anything quite like it and have spent hours just checking it out. The website deals specifically with representations of time, space, and globalization in corporate advertising and the information presented is truly awesome. You've got to see and read it to fully comprehend the implications of what's going on around us.

The Church of Stop Shopping

http://www.revbilly.com

OK—get ready for a wild ride! This is just an unbelievable site and I have got to wonder how effective this really is. I am not kidding. Check this guy out.

Adbusters

http://adbusters.org/home

One of my long-time favourites and I always like to keep this in my bookmarks to watch for new ads that they've created. The way this organization can make people think absolutely astounds me—as I hope it will you! Check out their photo gallery—brilliance at its best.

No Sweat

http://www.nosweatapparel.com/index.html

This link brings you to a site that uses union-made products—and cool ones at that—whose belief is that the only viable response to globalization is a global labour movement. You can also shop online. For shoes—and so much more—but who needs more when you have shoes?

Critical "Linking" Question

As if you had to think twice on this one. Scope out The Church of Stop Shopping and spend some time on that site—well, at least as much as you can handle. Exactly what do you think Reverend Billy is trying to accomplish? Do you really believe that his campaign will change coffee consumption or do you think he's promoting his own ideologies and beliefs?

Solutions

True or False?

1. F
2. F
3. T
4. F
5. T
6. F
7. T
8. T
9. F
10. F

Multiple Choice

1. B
2. C
3. A
4. B
5. D
6. B
7. B
8. C
9. B
10. A
11. D
12. D
13. B
14. D
15. B

Chapter 20
Research Methods

Chapter Introduction

Now that you're aware of what's going on around you and how it can affect your life, it's time to think about the ways in which sociologists gather the information that attempts to describe your world. This is very often easier said that done, because you're probably not used to thinking about the creation or determination of social factors. Just like any other scientists, social scientists attempt to discover evidence that can be used to explain and predict how or why our social realities exist. Although it's very different from the methods that are used in other sciences (either natural or social) social research is all about the purposeful, systematic, and rigorous collection of information. It's often difficult in sociology because we're trying to gather data that will help us to predict the behaviour of human beings, and as you probably already know, it's often hard to understand people, let alone predict what they're going to do next!

This chapter reveals some of the most popular methods of collecting meaningful information from and about people. Whether you've been reading your textbook, listening to your instructor or working with this guide, I really hope that you've been thinking "sociologically." Instead of just believing everything that you hear or read about, do you question where information comes from? If you read the newspapers, is one paper more credible than another? One of my very favourite things to do is to read about the same news item in three different newspapers. It's amazing how much the coverage of the event differs. Why do you think this is? Do you automatically believe the messages that television commercials are trying to get across? Have you ever wondered where those four out of five dentists come from or how huge corporations determine how many people prefer Coke instead of Pepsi? Often, it's sociologists that gather this information, and when you begin to critically evaluate the information that you hear or read, you'll get an idea of just how interesting the life of a social researcher can be.

Jump Start Your Brain!

In this chapter you will learn that:

1. Science is one of several sources of knowledge. Like other sources of knowledge, it can be wrong. However, unlike other ways of knowing, science uses methods of gathering theoretically relevant evidence that are designed to minimize error.
2. Research methods are used by sociologists to gather evidence in order to test theories about recurring patterns of human activity. Underlying these techniques is a variety of assumptions about the nature of facts, objectivity, and truth.
3. In comparison with the evidence available to natural scientists, an added complexity confronts social scientists: humans assign meaning to their actions, and interpreting meaningful action is very complicated.

4. Sociologists have devised many useful methods of obtaining evidence about the social world, including experiments, interviews, observational techniques, and surveys.
5. Good sociological research adds to our knowledge of the social world, expanding opportunities and options by helping to solve social problems.

Quiz Questions

True or False?

1. Values have the potential to bias or distort observations and both the natural and the social sciences must guard against distortion. If the scientific method is defined as a set of practices or procedures for testing the validity of knowledge claims, both chemists and sociologists could be seen as "doing" science. True or False
2. The most efficient way to prove causation in sociology is by showing the correlation between two or more variables. True or False
3. Among the rules of scientific method working to enhance objectivity and eliminate personal bias are critical scrutiny and full disclosure. True or False
4. To understand the meaning of social action requires being able, at least in principle, to participate in the social activity of which the action is a part. True or False
5. Only a scientific system of knowledge invariably produces truth, unerringly generating eternal accuracy. True or False
6. Sociologists have devised many useful methods of obtaining evidence about the social world, including experiments, interviews, observational techniques, and surveys. True or False
7. Social research involves systematically collecting theoretically relevant data in a way that minimizes error. It helps us to distinguish sociology from mere opinion. True or False
8. Science is founded on facts derived from direct observation. True or False
9. The Hawthorne effect refers to changes in people's behaviour caused by their awareness of being studied. True or False
10. The independent variable is a variable that is presumed to affect or influence other variables; it is the causal variable. True or False

Multiple Choice

1. "Men have higher annual incomes than women." In this statement, gender is the _____________ variable.

   a. independent

   b. dependent

c. confounding

d. mediating

2. _____________ are designed to reduce the likelihood that we are dealing in artefacts, while enhancing the likelihood that we have reproducible evidence.

a. Theories

b. Hypotheses

c. Research methods

d. Scientific facts

3. _____________ stresses that observations should be free of the distorting effects of a person's values and expectations.

a. Validity

b. Reliability

c. Consistency

d. Objectivity

4. _____________ refers to accuracy or relevancy.

a. Reliability

b. Internal consistency

c. Random selection

d. Validity

5. _____________ is a procedure used in experiments to assign test subjects to experimental conditions on the basis of chance.

a. Ethnography

b. Randomization

c. Objectivity

d. Sampling

6. _____________ is an incorrect inference about the causal relations between variables.

a. Causation

b. Spuriousness

c. Randomization

d. Confounding

7. The concept _____________ (making unconscious mistakes in classifying or selecting observations) is now commonly discussed as a danger to good methodological procedure.

a. selective data manipulation

b. observer bias

c. researcher prejudice

d. unconscious discrimination

8. By imagining yourself in the role of another, you come to appreciate someone else's point of view. This process, called "taking the role of the other," is reflected in the work of ____________.

a. Max Weber

b. George Herbert Mead

c. Erving Goffman

d. Gregor Mendel

9. Which of the following is *not* appropriate when designing questionnaires?

a. assuming the people understand what you arc asking

b. assuming that people know the answer to questions

c. assuming the people will give valid answers to questions

d. all of the above

10. We must be cautious in generalizing the results of laboratory experiments to non-laboratory situations. This concern is technically expressed as a problem of ____________ validity, or the degree to which experimental findings remain valid in non-laboratory situation.

a. face

b. external

c. internal

d. empirical

11. A ____________ is an unverified but testable knowledge claim about the social world.

a. hypothesis

b. theory

c. paradigm

d. statement

12. ____________ was one of the first sociologists to address the issue of interaction as a problem of social research. He argued that our interactions with other people draw upon meanings.

a. Robert Brym

b. Emile Durkheim

c. Karl Marx

d. Max Weber

13. According to the textbook, which is *not* a feature of social science research?

    a. Research results are confronted by the critical skepticism of other scientists.

    b. Social theory guides, either directly or indirectly, the evidence gathered.

    c. Social facts or findings are presented in a meaningful way.

    d. Evidence is systematically collected and analyzed.

14. David Hume disputed the popular argument of his day that science begins with observation. He argued that no matter how many observations that you make, you cannot infer that your next observation will be identical. This is known as _____________.

    a. deductive logic

    b. the problem of induction

    c. anomalous evidence

    d. trial and error

15. To help in summarizing numerical information, social scientists routinely rely on _____________.

    a. scientific formulas

    b. the work of other scientists

    c. theoretical interpretation

    d. statistical techniques

Critical Thinking

1. This chapter introduces the very important notion of "value-free" or unbiased research. If our perceptions of reality can be affected by our values, then how can scientists ever know for certain that what they see is true? Chapter 1 asked you to think about the suicide of Kurt Cobain. Now, imagine that you have been hired (and will be paid a lot of money) as a sociologist to determine who supports the "suicide" theory. Would your taste in music influence the research design that you would use for your study? Would your sample include fans, police officers, and/or musicians? How would you collect your data? Can you think of any personal biases or value judgments that might affect your method of inquiry? Can value free or unbiased research exist when studying humans?

2. In this chapter, the author asks what counts as scientific evidence and uses the criminal justice system as an example. In the cases of Guy Paul Morin and David Milgaard, the judges, juries, and prosecutors had all weighed evidence that they believed demonstrated the guilt of these men. Circumstantial evidence, filtered by expectations and values, had led justice astray. Subjective judgments seriously compromised these men's lives. It was only through DNA testing, or scientific

evidence, that these men were set free. Does the existence of DNA testing change your attitudes as they relate to capital punishment? Have television shows like *CSI* and *Cold Case Files* diminished your faith in the justice system? Are you likely to support capital punishment if an offender's guilt can be scientifically proven?

3. Without a doubt, the age of technology is upon us. The use of self-administered questionnaires has become a popular way of conducting interviews. Many of you have grown up in a world full of computers and have relied on them for school, work, and pleasure. Try to answer this question honestly: Have you ever completed an online survey, but pretended to be someone else? Have you ever used your pet's name or changed your age? What about submitting information to the online dating sites, or just checking out the people that do use these services? How much of the information you see on the Internet is true? How reliable and valid do you think this information is?

4. Brian Wilson (2002) used a number of different methods (a cluster of methods) to gather important information about the youth subculture in Canada. Not only did he become a participant observer by attending rave meetings and parties, he also interviewed people who were actually ravers. If you were asked to research the same group of ravers, what would you do differently? Do you think that your interview subjects would act differently because they knew that they were participating in a research study? How might you account for this when you report your results?

Time to Gear Down!

At the conclusion of this chapter, you should be able to discuss or write about the following, without having to rely on the textbook:

1. Research methods are ways of getting evidence to test suppositions about the world around us. Behind the various techniques (e.g., experiments, interviews) we use to obtain evidence and expand our knowledge of the social world, we must recognize important assumptions about such things as facts, objectivity, and truth.

2. Science is one of several sources of knowledge. Like other kinds of knowledge, scientific knowledge can be wrong. However, unlike other ways of knowing, science incorporates explicit methods designed to reduce error in what is currently accepted as scientific knowledge. Evidence must be systematically collected and rigorously evaluated.

3. Good science integrates both good theory and good research. The latter two are inseparable. Theories are ideas about how the world works or claims about how to explain or understand the recurring, patterned nature of human activity.

4. Evidence is crucial to developing, revising, or discarding theoretical claims. In comparison with the evidence available in the natural sciences, the evidence available to social scientists presents added complexity, because of the meaningful character of human social action. People, unlike molecules, assign meaning to their actions and the actions of others.

5. Sociologists have devised many useful methods for obtaining evidence about the social world. Observation and questioning are the two principal techniques, although each of them is conducted by using a variety of formats, including experiments, surveys, participant observation, and interviews.
6. Good research adds to our knowledge of the world around us. Such knowledge expands our opportunities and options. Sociological knowledge helps either in solving social problems or by sensitizing us to our collective human condition, expanding our social horizons.

Web Links

Internet for Social Research Methods

http://www.vts.rdn.ac.uk/tutorial/social-research-methods

This site offers a free and really valuable tutorial that can assist you in developing your Internet information skills.

Research Methods Tutorials

http://www.socialresearchmethods.net/tutorial/tutorial.htm

This website is very cool, because it was originally a project designed and written by graduate students for undergraduate students in the social sciences. It offers tutorials on specific topics related to research methods and presents them in a way that's easy to understand and fun to learn!

Research Methods in the Natural and Social Sciences

http://www.mcli.dist.maricopa.edu/proj/res_meth/index.html

A must-try website!!! This site is awesome—especially for learning the different research methods, because it allows you to enter a simulated lab and compare research settings and findings between disciplines. A totally excellent experience for visual and hands-on learners, this site will be a phenomenal study tool.

Canadian Sociology and Anthropology Association

http://www.csaa.ca/structure/Code.htm

Here's the link to the Statement of Professional Ethics of the Canadian Sociology and Anthropology Association. Take the time to look at all the hoops that researchers must jump through before they get their proposals approved.

Statistics Canada—List of Questionnaires

http://www.statcan.ca/english/sdds/indexti.htm

This is it—the motherlode! This is where you'll find all the questionnaires that Statistics Canada uses to collect their data—virtually every topic that you can think of. Check it out and see how valid and reliable some of their measures are, now that you're a research whiz!

Critical "Linking" Question

Hop on the Statistics Canada website and choose a questionnaire that you'd be really familiar with—say the one on movies. Go through it and see if you can verify that it is indeed valid and reliable.

Solutions

True or False?

1. T
2. F
3. T
4. T
5. F
6. T
7. T
8. F
9. T
10. T

Multiple Choice

1. A
2. C
3. D
4. D
5. B
6. B
7. .B
8. C
9. D
10. B
11. A
12. D
13. C
14. B
15. D

Chapter 21
Networks, Groups, Bureaucracies, and Societies

Chapter Introduction

So? What's your strategy for beating the beginning-of-term bureaucracy? Are you part of a brigade that takes turns standing in line so that you can spell each other off in case someone has to go to the bathroom or desperately needs a "Timmy's"? Do you silently repeat a calming mantra over and over again to stop your brain from literally bursting through your ears? Or are you one of the smarter ones, who can just face the music, realize that you'll spend countless hours in lines and frustration, and surrender to the insanity? When I was an undergraduate student at Brock—and remember, I typed my first thesis on a *typewriter*—they used to have registration in the gymnasium and the lines would snake all the way up the stairs and out into the parking lot. Literally, it would take seven or eight hours just to get to the front of the line, only to find that the course you had chosen was full, so you had to go back to the end of the line and start all over again. Though I will not divulge my secrets, let's just say that I quickly found ways around that one. It's funny, because in all other ways I have the patience of a saint, but I do not like to have my time wasted—especially by bureaucracy and red tape. And now, twenty years later, I find myself in the same situation as I register again each September for my doctoral work at OISE/UT. Though admittedly the process has improved, it is still quite frustrating.

But standing in those long lines did have their advantages. I would meet students that I hadn't seen all summer, and it was a great opportunity to meet people new to the campus and, even better, new to Canada. In this chapter, you'll learn about the importance of social networks and you'll be surprised at how important those brief meetings may end up being in your life. I think you'll be awed by the "six degrees of separation" stuff, and social network analysis presents some incredible research in its own right. My first master's thesis addressed nepotism in the workplace and indeed supported the claim that it's not those who are really close to you that can help you land that dream job, it's the people that you meet once in a while, but don't know very well that can sometimes "hook you up."

In this chapter you'll discover the wild world of bureaucracy, and get an understanding of why it sometimes seems as frustrating as it is. You'll also discover the logic behind group behaviour and see what sociologists study when they study groups by looking at network analysis. The author summarizes the chapter by applying these sections to society as a whole so that we can see how networks, groups, organizations, and entire societies can be mobilized for good or for evil!

Jump Start Your Brain!

In this chapter you will learn that:

1. We commonly explain the way people act in terms of their interests and emotions. However, sometimes people act against their interests and suppress their emotions because various social collectives (groups, networks, bureaucracies, and societies) exert a powerful influence on what people do.
2. We live in a surprisingly small world. Only a few social ties separate us from complete strangers.
3. The patterns of social ties through which emotional and material resources flow from social networks. Information, communicable diseases, social support, and other resources typically spread through social networks.
4. People who are bound by interaction and a common identity form social groups. Groups impose conformity on members and draw boundary lines between those who belong and those who do not.
5. Bureaucracies are large, impersonal organizations that operate with varying degrees of efficiency. Efficient bureaucracies keep hierarchy to a minimum, distribute decision making to all levels of the bureaucracy, and keep lines of communication open between different units of the bureaucracy.
6. Societies are collectives of interacting people who share a culture and a territory. As societies evolve, the relationship of humans to nature changes, with consequences for population size, the permanence of settlements, the specialization of work tasks, labour productivity, and social inequality.
7. Although our freedom is constrained by various social collectives, we can also use them to increase our freedom. Networks, groups, organizations, and entire societies can be mobilized for good or evil.

Quiz Questions

True or False?

1. The relationship between people and nature is the most basic determinant of how societies are structured and therefore how people's choices are constrained. True or False
2. In postindustrial societies, women have been recruited to the service sector in disproportionately large numbers, and that has helped to ensure a gradual increase in equality between women and men in terms of education, income and other indicators of rank. True or False
3. Much evidence suggests that flatter bureaucracies with decentralized decision making and multiple lines of communication produce more satisfied workers, happier clients, and bigger profits. True or False

4. Authoritarian leaders try to include all group members in the decision-making process, taking the best ideas from the group and moulding them into a strategy that all can identify with. True or False
5. The boundaries separating groups often seem unchangeable and even "natural." In general, however, dominant groups construct group boundaries in particular circumstances to further their goals. True or False
6. Conformity is an integral part of group life and secondary groups generate more pressure to conform than primary groups. True or False
7. Although intensity and intimacy characterize triadic relationships, outside forces often destroy them. True or False
8. One of the advantages of network analysis is its focus on people's actual social relationships rather than their abstract attributes, such as their age, gender, or occupation. True or False
9. According to Mark Granovetter (1973) mere acquaintances are more likely to provide useful information about employment opportunities than friends or family. True or False
10. As Max Weber (1978) defined the term, a *social structure* is a large, impersonal organization comprising many clearly defined positions arranged in a hierarchy. True or False

Multiple Choice

1. Until about 10,000 years ago, all people lived in ____________ societies.
   a. horticultural
   b. foraging
   c. industrial
   d. agricultural
2. The crystallization of the idea of ____________ was one of the most significant developments of the agricultural era.
   a. apprenticeship guilds
   b. communal living
   c. private property
   d. the barter system
3. Stimulated by international exploration, trade, and commerce, the Industrial Revolution began in ____________ in the ____________.
   a. France; 1650s
   b. Japan; 1800s
   c. Russia; 1720s
   d. Britain; 1780s

4. The invention of recombinant DNA marked the onset of a new social era—what we prefer to call the era of ___________ society.

   a. postindustrial

   b. postnatural

   c. neonatural

   d. present day

5. Bureaucracies are large, impersonal organizations that operate within varying degrees of efficiency. Which of the following is *not* a characteristic of an efficient bureaucracy?

   a. keeps hierarchy to a minimum

   b. distributes decision making to all levels of the bureaucracy

   c. keeps lines of communication open to all levels of the bureaucracy

   d. ensures both vertical and lateral mobility of all workers within the bureaucracy

6. As Milgram's experiment illustrates, structures of authority tend to render people obedient because. Most people find it difficult to disobey authority because of fear. Which of the following is *not* one of the things they fear?

   a. ridicule

   b. accountability

   c. ostracism

   d. punishment

7. This sociologist contrasted "community" with "society," and according to him, a community is marked by intimate and emotionally intense social ties, whereas a society is marked by impersonal relationships held together largely by self-interest.

   a. Mark Granovetter

   b. Frederick Tonnies

   c. Max Weber

   d. Robert Michels

8. Beginning in the early 1970s, such corporations as Volvo and Totoyta were at the forefront of bureaucratic innovation and began eliminating ____________ positions.

   a. CEO

   b. secretarial

   c. middle-management

   d. vice-president

9. Several aspects of the organizational environment help to explain Japanese-American differences in the 1970s. Which of the following is *not* one of them?

   a. There was a stronger Japanese worth ethic, based in their cultural teachings.

   b. Japanese workers were in a position to demand and achieve more decision-making authority than American workers were.

   c. International competition encouraged bureaucratic efficiency in Japan.

   d. The availability of external suppliers allowed Japanese firms to remain lean.

10. Research shows that thc least effective leader is the one who allows subordinates to work things out largely on their own, with almost no direction from above. This is known as ____________.

    a. the absent mediator approach

    b. the employee restoration model

    c. the conflict resolution module

    d. the laissez-faire leadership style

11. Traditionally, sociologists have lodged four main criticisms against bureaucracies. Which of the following is *not* one of them?

    a. bureaucratic ritualism

    b. efficiency vs. effectiveness

    c. dehumanization

    d. bureaucratic inertia

12. Sociologists make a basic distinction between primary and secondary groups. Which of the following is *not* a characteristic of a primary group?

    a. Norms, roles, and statuses are agreed on but are not put in writing.

    b. Social interaction creates strong emotional ties.

    c. The group extends over short periods of time.

    d. It results in group members knowing one another well.

13. There are more than 32 million Canadians, of whom just a few hundred are your family members, friends, acquaintances, and work colleagues. Yet, if you were asked to get in touch with a complete stranger on the other side of the country by using only personal ties, it would take only about ____________ contacts to reach the stranger.

    a. 4

    b. 6

    c. 8

    d. 12

14. Milgram's experiment teaches us that as soon as we are introduced to a structure of authority, we are inclined to obey ____________.

    a. those in power

    b. our first instincts

    c. a code of ethics

    d. the rules of the game

15. A study of Nazis who roamed the Polish countryside to shoot and kill Jews and other "enemies" of Nazi Germany found that the soldiers often did not hate the people they systematically slaughtered, nor did they have many qualms about their actions. They simply developed deep loyalty to each other. It is the power of ____________ that helps us understand how soldiers are able to undertake many unpalatable actions.

    a. camaraderie

    b. laws of deviance

    c. rules of conformity

    d. norms of solidarity

Critical Thinking

1. Is it group loyalty or is it betrayal? Whatever it is, in my mind, it is unacceptable. Read Box 21.1 on page 21-5. What is your first reaction after reading this? I can't decide if mine was disbelief or anger that initially struck me. If in fact this was group cohesion, how do you think the members of the group would have rewarded themselves for their loyalty? What do you think happened to one member that betrayed the group? This incident occurred in 1989—do you think it would happen today?

2. Princeton University biologist Lee Silver and Nobel Prize–winning physicist Freeman Dyson go so far as to speculate that the ultimate result of genetic engineering will be several distinct human species and that people who are in a position to take full advantage of it will be better-looking, more intelligent, less likely to suffer from disease, and more athletic. Others will not be so lucky and will have their fate determined by "true democracy." This seems like a very ominous prediction to me—do you have the same reaction? Given the chance, would you genetically-engineer your baby? At the very least, would you predetermine your child's gender if you could?

3. According to the work of Mark Granovetter (1973), you are likely to find a job faster if you understand the "strength of weak ties" in microstructural settings. This is very contrary to common sense, because everyone thinks that you have to "know someone" to get a really good job. Do you agree with his findings? Do you know people who got great jobs because of their "connections"? How helpful is it to "know someone" when it comes to finding meaningful work? After reading about his work on network analysis, do you still believe that "it's not what you know but who you know" that counts?

4. I always make up these stupid games to play when I'm waiting with someone for an appointment or we're just hanging out, and I ask questions like "If you could do any sport on Wide World of Sports (I think it's ESPN now), what sport would you pick to do?" So this question has you looking at Table 21.1 on page 21-21 and checking out the transformation of human societies over the past 100,000 years. In which society do you think you'd stand the best chance of survival? Personally, I know there is no way I could survive until at least the agricultural societies came around. I don't do foraging very well. What do you think would be the most challenging aspects of each societal type for you?

Time to Gear Down!

At the conclusion of this chapter, you should be able to discuss or write about the following without having to rely on the textbook:

1. People's motives are important determinants of their actions, but social collectives also influence the way they behave. Because of the power of social collectives, people sometimes act against their interests, values, and emotions.
2. It is a small world. Most people interact repeatedly with a small circle of family members, friends, coworkers, and other strong ties. However, our personal networks overlap with other social networks, which is why only a few links separate us from complete strangers.
3. Network analysis is the study of the concrete social relations linking people. By focusing on concrete ties, network analysts often come up with surprising results. For example, network analysis has demonstrated the strength of weak ties in job searches, explained patterns in the flow of information and communicable diseases, and demonstrated that a rich web of social affiliations underlies urban life.
4. Groups are clusters of people who identity with one another. Primary groups involve intense, intimate, enduring relations, secondary groups involve less personal and intense ties, and reference groups are groups against which people measure their situation or conduct. Groups impose conformity on members and seek to exclude nonmembers.
5. Although bureaucracies often suffer from various forms of inefficiency, they are generally efficient compared with other organizational forms. Bureaucratic inefficiency increases with size and degree of hierarchy. By flattening bureaucratic structures, decentralizing decision-making authority, and opening lines of communication between bureaucratic units, efficiency can often be improved.
6. Social networks underlie the chain of command in all bureaucracies and affect their operation. Weber ignored this aspect of bureaucracy. He also downplayed the importance of leadership in the functioning of bureaucracy. However, research shows that democratic leadership improves the efficiency of bureaucratic operations in noncrisis situations, authoritarian leadership works best in crises,

and laissez-faire leadership is the least effective form of leadership in all situations.

7. The organizational environment influences the degree to which bureaucratic efficiency can be achieved. For example, bureaucracies are less hierarchical where workers are more powerful, competition with other bureaucracies is high, and external sources of supply are available.
8. Over the past 100,000 years, growing human domination of nature has increased the supply and dependability of food and finished goods, productivity, the division of labour, and the size and permanence of human settlements. Class and gender inequality increased until the nineteenth century and then began to decline. Class inequality began to increase in some societies in the last decades of the twentieth century and may continue to increase in the future. In foraging societies, people lived by searching for wild plants and hunting wild animals. Horticultural and pastoral societies emerged about 10,000 years ago. In horticultural societies, people domesticated plants and use simple hand tools to garden. In pastoral societies, people domesticated cattle, camels, pigs, goats, sheep, horses, and reindeer. Agricultural societies first emerged about 5000 years ago. In such societies, people used ploughs and animal power to produce food. Great Britain was the first society to industrialize, beginning about 225 years ago. Industrial societies used machines and fuel to greatly increase the supply and dependability of food and finished goods. Shortly after World War II, the United States became the first postindustrial society. In postindustrial societies, most workers are employed in the service sector and computers spur substantial increases in the division of labour and productivity. Some societies may be said to havc entered a postnatural phase in the early 1970s, when genetic engineering became possible. Genetic engineering enables people to create new life forms, holding out much promise for improving productivity, feeding the poor, and ridding the world of disease, and so on, and much uncertainty as to whether these benefits will be equitably distributed.
9. Networks, groups, bureaucracies, and societies influence and constrain everyone. However, people can also use these social collectives to increase their freedom. In this sense, social collectives are a source of both constraint and freedom.

Web Links

Small World Project

http://smallworld.columbia.edu

As discussed in Box 21.2, the "six degrees of separation" idea is now a project run by Columbia University's Department of Sociology. Visit this website to become part of the project and experience firsthand your own degree of separation.

The *Wall Street Journal* Executive Career Site

http://www.careerjournal.com/jobhunting/networking

Get ready to schmooze! This site provides various articles on the success and importance of networking when you're looking to make those important connections that will land you that job!

Bureaucracy by Honoré de Balzac

http://www.literature.org/authors/de-balzac-honore/bureaucracy/index.html

De Balzac was a highly influential and eccentric (he was totally addicted to coffee) author in the 1800s in Paris—he wrote about everything that reflected social life at the time. This link is a rare treat indeed—an online book that he wrote about his take on bureaucracy!

Bystander Apathy

http://teenadvice.about.com/library/weekly/aa121302a.htm

This site gives you more information about bystander apathy, but even more importantly, offers tips on street safety too. Check out the Street Smart Quiz!

The Human Genome Project

http://www.ornl.gov/sci/techresources/Human_Genome/home.shtml

OK—trust me, this site has everything you ever wanted to know—and more—about DNA and DNA analysis. Very, very controversial stuff happening here.

Critical "Linking" Question

Are you up for the challenge? I double-dare you to read de Balzac—I guarantee you, it will be worth it. Can you find commonalities between his notions of bureaucracy in 1800 Paris and bureaucracy today?

Solutions

True or False?

1. T
2. F
3. T
4. F
5. T
6. F
7. F
8. T
9. T
10. F

Multiple Choice

1. B
2. C
3. D
4. B
5. D
6. A
7. B
8. C
9. A
10. D

11. B
12. C
13. B
14. A
15. D